Praise for
Table Talk Math

"This is an exciting time in math education. There is practically no end to the number of resources designed to help teachers (and parents!) spark rich mathematical conversations. Yet this vast landscape can also be intimidating. Where should we begin when there is so much to choose from? What questions can we ask our students (and children) to prompt delight-filled discussions? And how should we respond to *their* questions and ideas in return? In *Table Talk Math*, John Stevens offers timely answers to these questions and more. If searching on our own through everything the Internet has to offer is like drinking water from a fire hose, then *Table Talk Math* is John serving it up one cool, refreshing glass at a time."

—Michael Fenton, reasonandwonder.com

"Time spent with our children is limited, leaving us scrambling for more meaningful ways to build life-long relationships with them. As parents, we want to make the most of every opportunity, but it's tough. It's here where *Table Talk Math* offers simple ways for us to purposefully connect with the ones we love, but in a mathy kind of way. Through the eyes of his own childhood, John Stevens shares the personal stories that helped foster his own curiosity in the world around him and he invites you to do the same—just as his parents did for him."

—Graham Fletcher, gfletchy.com

"*Table Talk Math* is chock-full of stories that will make you feel like you're seated at John's table. This practical guide provides the perfect balance of accessible resources, empowering coaching, and authentic humor. Teachers and parents, get ready to pull up a chair and invite students of all ages to join mathematical conversations that develop number sense, perseverance, and joy."

—Cathy Yenca, mathycathy.com

"As an elementary school principal, I'll do everything I can to get *Table Talk Math* in the hands of parent communities. John Stevens' honest experiences about the power of having meaningful and intentional conversations with kids is truly transformative. If you're looking for ways to promote curiosity, confidence, and risk-taking in your own children, this book is, without a doubt, for *you*!"

—Amy Fadeji, mrsfadeji.blogspot.com

"John Stevens is much more than a math teacher; he's a storyteller. In *Table Talk Math*, he invites us into his home and shows us, through personal vignettes with his two children, how we can work math into our own everyday conversations. Enhanced with contributions from a team of math innovators and written in a refreshingly readable style, *Table Talk Math* give parents easy ideas for bringing math home in authentic and relevant ways."

—Jennifer Gonzalez, editor-in-chief at
Cult of Pedagogy, cultofpedagogy.com

"How appropriate that a book designed to help parents have conversations about math with their children is written in such a conversational format. If more authors approached math in the way that John Stevens does in *Table Talk Math*, we might be able to begin to erase the stigma that surrounds math in our culture."

—Donna Boucher, math consultant and coauthor of
Guided Math Workshop, mathcoachscorner.com

"John Stevens works harder than anyone I know. In this book, he's scoured the Internet and connected with real-life math teachers to find accessible, understandable, tangible ways for parents to support their students in math, the least-accessible subject for many parents. In *Table Talk Math*, parents will find tips and tricks, ideas that we wish all of our students' parents would use. Stevens is making math accessible for parents."

—Matt Vaudrey, mrvaudrey.com

"Parents know the value of reading with their children from a young age. *Table Talk Math* not only brings to light the importance of doing math with children but also teaches exactly how to do it! John shares a multitude of strategies and diverse resources that appeal to children of all ages, from Pre-K through high school. These fascinating resources will make exploring math with your children the most highly anticipated part of your dinner routine! Avoid math anxiety and begin fostering a love of mathematics in your children today!"

—Julie Reulbach, ispeakmath.org

"You might see the title *Table Talk Math* and think, 'Oh, a book to help my kid get better at math.' You'd be mostly wrong—and kinda sorta right. This isn't a book about teaching your kids math so they can excel beyond their peers. Rather, John Stevens' purpose is to give parents an avenue and resources for building relationships with their children by being curious about mathematics and the world around them. Sure, rich, engaging conversations about mathematics will likely help your child understand math better so that they can do better at school, but that's a secondary effect. More importantly, these conversations will become a special part of your day. They'll become a way for you and your child to connect with one another and enjoy the time you have together, especially if your time is more limited than you would like.

"As a parent, I appreciate John's down-to-earth writing style. Throughout the book he weaves in stories from his own childhood as well as from his time as a parent and teacher. It lends an authenticity to his message; he's lived his message. I also appreciate how he invites other experts into the book. While this book might be the spark, the fuel for the fire is the wealth of free resources shared across the chapters. Not only do you get to hear about how John has used them, but he invites the creators of these resources to share their experiences and insights. You should definitely have a web browser handy while you read the book, because you're going to want to stop regularly to visit all the websites he shares!

"As an educator, I appreciate how John demonstrates how accessible mathematical conversations are for children of all ages. He does a fantastic job of modeling how conversations can be tailored depending on the age of your child. I'm also a fan of the section at the end of each chapter where he gives concise advice about what he encourages in your conversations with your children and what you should definitely avoid.

"In the end, what I love most about this book is its message: Math is for everyone—parents and children alike—and it's even better when it's enjoyed together."

—Brian Bushart, bstockus.wordpress.com

T
TABLE
L
K MATH

A Practical Guide to Bringing
Math into Everyday Conversations

John Stevens

Foreword by Christopher Danielson, PhD

Table Talk Math
© 2017 John Stevens

This book is available at special discounts when purchased in quantity for use as premiums, promotions, fundraising, and educational use. For inquiries and details, contact us: shelley@daveburgessconsulting.com.

Published by Dave Burgess Consulting, Inc.
San Diego, CA
http://daveburgessconsulting.com

Cover Design by Genesis Kohler
Editing and Interior Design by My Writers' Connection

Library of Congress Control Number: 2017935672
Paperback ISBN: 978-1-946444-02-8
Ebook ISBN: 978-1-946444-03-5

Contents

Foreword xi

Introduction xiii

Chapter 1: "The Talk" 1

Chapter 2: Less is More—More or Less 5

Chapter 3: Debates 13

Chapter 4: Estimations. 25

Chapter 5: Patterns 39

Chapter 6: Organization 55

Chapter 7: Fraction Talks. 67

Chapter 8: Talking About Math 79

Chapter 9: Noticing and Wondering 85

Chapter 10: Leveraging Technology... or Not. 95

Chapter 11: Conclusion 101

Additional Resources. 103

Contributors 106

Acknowledgments 108

More From Dave Burgess Consulting, Inc. 110

About the Author 119

Dedication

For my boys, Luke and Nolan

Foreword

Children have powerful minds, and they delight in using them. The culture in which children grow up shapes their minds. We know that when children grow up in homes with many books they are more likely to be readers. We know that the number of words children hear before entering school directly impacts their vocabulary. In areas from reading to sports to art, we understand that parents play an important role in supporting their children's development.

But we tend to talk about math differently. We separate people—children even—into two categories. These are the math people; those people are not math people.

People talk about math this way because of common, but incorrect, ideas about math learning. A growing body of research demonstrates that children learn math in very nearly the same ways they learn language. It turns out that supporting children's mathematics learning is very much like supporting their learning to read.

Children learn math from the culture by which they are surrounded. When their parents support and encourage talk about numbers, shapes, and patterns, children tend to talk more about those things. And through talking about them they learn. When their parents are anxious about math, or talk about themselves as not being math people, children come to understand that math is only for some people, and that it is scary and probably best avoided.

I had years of teaching experience before having children. As a teacher and curriculum developer, I had come to understand that most of the math children learn is in school. I thought school was responsible for drawing children's attention to multi-digit numbers, geometric shapes, and algebraic patterns. When my daughter was four years old, she shattered that understanding in a single conversation by asking me, "Why don't circles have tips?" In a five-minute conversation, I realized that young children don't just name circles, they think about the properties of circles. Math doesn't only arise in lessons. It is frequently on the minds of children. Math—along with language and play and body movement—is one of the ways children make sense of their world from an early age, even prior to schooling.

Table Talk Math acknowledges that children have ideas that need nurturing, and the book helps parents build a math-supportive home culture for their children. John Stevens and the contributors to *Table Talk Math* are not just math teachers who love math. As teachers, they stand out because of their love for the ideas their students bring to their classrooms—ideas these students have explored *outside* of school. These teachers understand that children encounter mathematical ideas in their everyday lives and they are experts at bringing those ideas alive in their classrooms. More importantly perhaps, these teachers are parents. They have watched at close range the mathematical development of their children, and they have made conscious decisions about how to support this learning. In *Table Talk Math*, they share these decisions with parents and help them to talk about numbers, patterns, and shapes in ways that will feel natural.

My hope is that the ideas of *Table Talk Math* will find a wide audience and help to create a culture of math learning for all children. More specifically, I hope you—the reader—will enjoy this book and will use these ideas for fun and learning at the table, on the couch, at the grocery store, or at the bus stop. Together we can support children's math learning in and out of school so that all kids can use math as they grow and participate in all aspects of our democratic society. In careers, domestic life, and public discourse, math matters. And we know that math starts at home.

Christopher Danielson
Teaching Faculty at Desmos, Inc. and
author of *Which One Doesn't Belong?*

Introduction

I'll never forget family hockey night—January 30, 1991, 7:22 p.m.

"Gretzky takes the puck across center ice, weaves in and out of traffic, dishes it back to Robitaille. Robitaille dumps it back to Granato, avoiding a big hit by Daneyko, leaving the puck for Watters ..."

Sitting in the living room on the carpeted floor of our double-wide home in northern Nevada, my family and I watched the Los Angeles Kings play yet another intense hockey game. Sportscasters Bob Miller and Nick Nickson took us through every pass, every check—all the subtle moves a hockey player makes over the course of a shift—in a way only they could. To us sports fans, those announcers were legendary. They made us feel a part of every moment of action.

"... Watters winds up, SHOOTS ... AND HE—"

Every moment—that is—until the power went off.

Without a flicker of warning—or the decency to wait until we saw the result of Tim Watters' slap shot—the power shut down. Again. It had been happening a lot lately, and there was nothing we could do. Living in a small mining town required some compromises; apparently doing without electricity was one of them. But I remember this particular outage like it was yesterday—not because we missed the end of the game—but because of what happened next.

My brother and I loved family hockey night because we could stay up an extra thirty minutes past bedtime, depending on the score of the game and whether or not we started arguing. After all, Mom and Dad were going to watch the game anyway, and my brother and I wouldn't be able to sleep with them coaching the team through the TV. When the power went out, though, instead of sending us on to bed, our parents knew exactly what to do—almost as if they had planned it.

Mom fumbled through the junk drawer in the kitchen,[1] and Dad went to the coat closet in the hallway. We knew what that meant: game night! Some elementary-age kids despised playing games with their parents, but we absolutely loved it. My brother, Matthew, who was six at the time, was old enough not to screw up the board but not quite old enough to hold his own during a game. He was also smart—smart enough to know who to upset and how to get away with it. Mom came back with the flashlights, and Dad came back with our new entertainment: *Othello*.

My dad once told me he and his college roommates would play Othello for hours—often in silence—as they worked through multiple moves and developed strategies.[2] Our family had another way of playing—no one was silent. Regardless, we had fun and without realizing it, playing the game taught us the essence of problem-solving and preparation.

Othello is a board game requiring a bit of skill and a *lot* of strategy. Here's the basic premise:

- The game is played with a set of colored chips: black on one side, white on the other.
- One team (or player) is black, the other is white.
- Taking turns, players place the chips on a board, starting from the center.
- When a player has bookended a strand of the opposing player's chips, all the chips in the strand are flipped to the bookend's color.

---

[1] Come on. We all have a junk drawer, don't we? Where the dead pens, used stamps, 76 cents in coins, and all our flashlights are stored? Yes—*all* of them.

[2] That was definitely *not* my experience in college.

My brother and I loved to play Othello because we got to flip over each other's game pieces. And we relished any chance we got to "legally" make each other angry.

We even forgot about the hockey game for a little while. (Remember, this was before the age of smartphones and the tether of 24/7 Internet access.) We found out the next morning that the Los Angeles Kings had lost to the New Jersey Devils. But that was okay because we had a great time playing, talking, and laughing together.

Maybe because my parents were both engineers (Dad mining and Mom mechanical).

Maybe because we grew up without a lot of money.

Maybe because we were raised in a small town.

Maybe because we moved a lot.

Maybe because we didn't have extended family anywhere close to us.

Maybe because my parents valued the time they had with their families growing up.

Whatever my parents' reasons for teaching us to appreciate time together, I am forever grateful. They seized any opportunity to engage us and challenge us to think about the situations we were placed in. And they always emphasized conversation.

As you read this book, you'll see how math conversations often arose (seemingly) spontaneously in our house. Those conversations shaped me as a child and continue to impact me as a parent and educator. From them, I learned two things: my parents liked to talk with me (not just at me), and no matter what I was going to be when I grew up, understanding math would make almost everything easier. Sure, there were the rough years where it wasn't cool to admit publicly I liked spending time with and actually talking to my parents. And there were years that sports and activities consumed most of our waking moments and made it difficult to find time for conversation.

But somehow my parents found ways to ask us questions not directly related to our school work.[3]

The night the power went out, Mom and Dad turned an unfortunate situation into an opportunity for the family to come together and work as a unit and to share thoughts and challenge one another. The whole time we were flipping over our chips and strategizing our next moves, we were employing math as a means to calculate the ideal setup for a winning game. It wasn't a worksheet or a set of problems that tightened the bond between us; it was the genuine joy of coming together and sharing what we truly enjoyed: each other's time.

That is why I chose to write *Table Talk Math*.

I want other parents to feel comfortable initiating a meaningful experience with their children. No matter what your mathematical background is, you—the parents or guardians of the incredible human(s) you are responsible for—can make a true difference in their perception of education and, in this specific case, mathematical awareness and fluency. In fact, society is counting on you to do that. My hope is that this book will help you be successful in that goal.

The National Council for Teachers of Mathematics (NCTM), the California Math Council (CMC), Twitter Math Camp (TMC), and other organizations have done a tremendous job of identifying the benefits of engaging in math-based discussions with children. These professional groups provide countless case studies, blog posts, journals, and anecdotes to elaborate on this topic, so I won't make you wade through a lot of research references. If you are interested in the research that explains the empirical value of math-based conversations with your children, I encourage you to seek out these resources.[4] This book, however, focuses on real-life experiences, stories that I hope you can relate to. In these pages, you'll find ideas you can try with your own

---

[3]   How many times have you asked your children, "How was school today?" or "What did you learn at school today?" Neither of these are bad questions, but we should find ways to engage our children when they get home from school beyond simply asking them to reflect on the seven hours they *just* spent sitting and being talked to/at/with.

[4]   Many of which have been curated by Dr. Jo Boaler and her team at youcubed.org.

household, and you may find solace in knowing you are not alone in your effort to be the best possible supporter, advocate, and parent for your little one(s).

Many of the examples I offer for initiating math-based conversations include stories from various stages in my life or the lives of my children, along with vignettes from creators of powerful websites that are sure to assist with ideas. I hope they will give you the confidence to try similar ideas and prompts with your child, no matter his or her age or grade level. Even our geometry, pre-calculus, AP statistics, and AP calculus students will find incredible value in fine-tuning their number sense. In fact, I've enjoyed many conversations with my friends and colleagues where math provided the ultimate solution. My point is that math-based conversations can benefit (and be fun for) everyone at the table—no matter how young or old. Yes, there are websites and areas of focus directly related to understanding upper-level mathematics for students and parents who wish to take that plunge. This book, however, is about making math accessible for everyone at the dinner table because math is for *everyone*.

Big kids can count in their head, and little kids can count on their fingers.

Big kids can use visuals, and little kids can build towers.

Big kids can play matching games, and little kids can critique the reasoning of others.

Big kids can make use of structure, and little kids can use structures to make sense.

Big kids can estimate, and little kids can model.

This book, however, is about making math accessible for everyone at the dinner table because math is for *everyone*.

Big kids can practice counting, and little kids can use math to justify a debate.

When we read an article, a blog post, or a book, the most natural inclination is to read it in context and determine whether or not the suggested advice would work for us and our child, or if it's just "not something my kid can do right now." That's why I didn't set up the chapters of this book based on age or grade level. I want you to feel free to adapt the prompts to suit your child's learning stage. While a conversation with a six-year-old will naturally be different than a conversation with a sixteen-year-old in depth and in detail, the prompts provided in *Table Talk Math* are not exclusive to one age group or another.

When is the last time that you and your teenage child sat down and played Monopoly, a game of Memory, or had a counting battle?

When is the last time that you and your elementary-aged child looked at a histogram from the news, an online source, or a study, and tried to make sense of it?[5]

I want to help show you how to get them talking—about math.

We can *all* have a voice at the table.

---

[5] Especially because it's so foreign to them, I love to sit down with my kids, show them a histogram representing different information, and try to make sense of it in front of them. Right now, they are four and six, but they are already engaged in trying to figure out what all of it means.

1

"The Talk"

Before we get into the fun stuff, let's sit down at the table—just you and me—and have a brief, but very important talk.

There is a lot of confusion and frustration about what is happening in education today. While some schools in the United States are implementing the Common Core State Standards, others are not. Some states want more autonomy; others do not. We parents want what's best for our kids, and that desire can make our emotions run high. In those moments, it's common to head to Facebook to vent, or maybe to YouTube to seek comfort from an ill-informed but somewhat relatable video posted by another frustrated parent. If we aren't consulting social media, we're looking to people who can validate our concerns.

Table Talk Math is *not* about Common Core. In fact, you won't see those two capitalized words written adjacent to each other again in this book. While there are parts about the standards I really like, most of them have no direct impact on what my students—and most importantly my own children—will need to be successful contributors to society. One major positive I have seen come from the shift has been the Standards for Mathematical Practice (SMPs). Here they are:[1]

- Make sense of problems and persevere in solving them.
- Reason abstractly and quantitatively.
- Construct viable arguments and critique the reasoning of others.

---

[1] "Standards for Mathematical Practice," *Common Core State Standards Initiative*, accessed March 24, 2017, http://www.corestandards.org/Math/Practice/.

- Model with mathematics.
- Use appropriate tools strategically.
- Attend to precision.
- Look for and make use of structure.
- Look for and express regularity in repeated reasoning.

Looking at those standards, think about the types of conversations you would hope that your child could enjoy. Within each of the SMPs are opportunities for your child to express their ideas, opinions, and estimates as well as opportunities to explore their educational curiosity and tinker with ideas until they make sense.

This book is also *not* about changing our current educational system. Large government and political structures are in place, as you may already know, making sweeping change very difficult. Sure, you could argue the ideas presented here *could* be used to make a large-scale difference in the infrastructure of education. But big picture reform isn't what this book is about.

Instead, *Table Talk Math is* about the conversations happening at the dinner table, the classroom table, a coffee table—or anywhere else for that matter. It's about intentionally and creatively engaging with your child even when busy schedules and life stressors make it challenging. Dual-income families and single parents are increasing at an alarming rate,[2] requiring many people to spend long hours away from their children each day, demanding more from them as parents *and* professionals. Even parents who have the opportunity to stay at home and raise their child face obstacles. For example, the pressure to pursue post high school education forces kids to mature faster, take on more, and explore less. If you have a college bound high-schooler, you're certainly aware of the pressure that kids face to boost their résumés to appeal to their favorite colleges and careers. As a result, they seem to have less time to be kids and to build meaningful relationships at home.

---

[2]   We have gone from 25 percent dual income families in 1960 to 60 percent in 2010. Source: "The Rise in Dual-Income Households," *Pew Research Center*, June 18, 2015, accessed March 24, 2017, http://www.pewresearch.org/ft_dual-income-households-1960-2012-2/.

My hope is this book will remind you—the caregiver of the precious human being sitting across from you at the dinner table, the provider who tirelessly and relentlessly creates growth opportunities for your child, the loving guardian who influences your child in ways no teacher, no educational system, no television, and no app can do—that *you* are the one who will make the difference in your child's life.

I've included a number of personal stories in this book, from my childhood and from my own ongoing adventure as a father. Sharing these stories prompted me to reflect on how I can improve as a parent—a worthwhile exercise. I encourage you to reflect on your own parenting experiences as you read this book and consider how you can best navigate your interactions with your child.

In addition to my insights, you will also hear from other teachers, parents, and community members who have dedicated their work in mathematics to the way children interact with numeracy. From these experienced educators, you'll discover that effective learning starts when we create an environment where encouragement and perseverance are the norm. This is often accomplished simply through intentional conversation. Intentionality is the key, especially in an era where busy lifestyles are the norm, stress is at an all-time high, and it just doesn't seem like the roses we pass by have much of an aroma anymore. After all, dinner used to be the most coveted time of day for family engagement, but seems to be a less common experience nowadays.

In 2003, Gallup ran a poll[3] of families who ate dinner together and found that only 28 percent of the American families surveyed ate dinner together seven nights a week.[4] Of those surveyed, 24 percent ate dinner together three or fewer nights per week. As sad as the statistic is, it is a reminder that we need to cherish the time we *do* have with our children and know that every interaction is a chance to make a difference. We can't always make our own schedule, work the day shift,

---

[3] Kiefer, Heather Mason, "Empty Seats: Fewer Families Eat Together," *Gallup*, January 20, 2004, accessed March 24, 2017, http://www.gallup.com/poll/10336/empty-seats-fewer-families-eat-together.aspx.

[4] Canada is at 40 percent and Great Britain is at 38 percent.

or fight traffic in time for a hot supper, but we *can* do our best to make the most of the time we have with our children.

Here are a few questions to ponder as you go through your day looking for the next engaging moment:

- How could the conversation be improved?
- What questions could I have asked to initiate a discussion when there was silence during [fill in any activity] today?
- If I approached my conversation in a different way, how might the result change?
- What did I see today at work, on my way home, or during lunch that might interest my child?
- When I see my child today, what is an interesting question I can ask?

Finally, *Table Talk Math* is not a one-stop-shop for solutions on parenting, nor something I intend to flaunt as the right way to parent.[5] Rather, my hope is to start a movement where it is safe and comfortable to engage in meaningful mathematics conversations around the dinner table, in the car on the way to sports practice, or hanging out in the backyard on a warm and sunny day.[6] If you are on Twitter, Instagram, or Facebook, keep the conversations going with the hashtag *#TableTalkMath*. I'll watch for great posts and put them up on the website:

tabletalkmath.com

Now... let's talk math.

---

[5] Trust me, I'm still working on the Parenting 101 On the Fly handbook, scrambling to find the best way to parent while keeping my sanity. If you have ideas, I'm all ears.

[6] Clarification: good conversations can happen when it's freezing cold, raining, snowing, windy—or any other weather condition. But really, I just love thinking about a warm, sunny day.

2

Less is More–More or Less

My children, Luke (six) and Nolan (four), must be on to me and my devious ways. I mean really, how could they *not* know they are the subject of a life-long experiment? Kids are smarter than we give them credit for. Think about how close they can get to stepping over "the line" without actually crossing it. Think about how your daughter knows you'll read one more book before bedtime or how your son always wears the shirt you *really* don't like.[1] Children are aware of and can expertly navigate the intricacies of the parent-child balance. But our kids don't care (much) about what schools we've attended, the jobs we've held, or the struggles we've endured. As such, it doesn't matter whether you hold multiple degrees or didn't make it through high school; you have something to offer your child in the form of encouragement and empowerment in math.

As we begin, I want to challenge you to be careful how you define *encouragement*. So many parents attempt to relate to their kids by sharing their own weaknesses, which quickly turns into parents being complicit in perpetuating a fear of math. So, when a child struggles with math, his mom or dad might say something like, "I know it's hard, I'm bad at math, myself!" Please, do your child a favor and remove "I'm bad at math" from your lexicon. An article in *The Washington Post* explains that, while parents say this as an attempt to connect with

---

[1] It's probably filthy, has been worn for eight days straight, and has a few extra LEGO pieces and Nerf darts in it. Oh, you as well?! Welcome to my life.

their children, the phrase does more harm than good.[2] Rather than comforting kids, that simple, over-used statement increases their math anxiety.

Our kids don't need more stress in a world where everything is overwhelming. There is an inherent challenge as children get older and meet their match with content, especially when it comes to the field of mathematics. For many students, this time is in middle school when the problems become more abstract, the arithmetic is difficult to do in our heads or on our fingers, and the pressure of performance is exacerbated by standardized testing and the looming high school graduation requirements. Signing off on a child's frustration may give parents a moment to commiserate and reminisce about their own struggles, but it does nothing to help and encourage the student—your child—persevere through this difficult time. And no, we don't have to make this complicated. Our kids should be enjoying life *and* learning. We owe it to them to provide the best possible experience we can, regardless of the paths we have taken to get here.[3]

Start the Conversation

Keeping conversations with children simple is important. Due to a number of factors, people have an ever-shrinking attention span. At least one survey[4] has found our ability to focus is limited to an average of eight seconds—less than that of a goldfish.[5] Kids most certainly *can* focus for longer than eight seconds. However, *expecting* them to focus for longer is sure to end in arguments and a bad taste for sitting around

---

[2]   Strauss, Valerie, "Stop telling kids you're bad at math. You are spreading math anxiety 'like a virus.'" *Washington Post*, April 25, 2016, accessed March 24, 2017, http://wapo.st/1Wma3Nd?tid.

[3]   As a fair disclaimer, my degree is in mathematics. I try to keep that train of thought *out* of conversations with my kids and work to initiate conversations at their level of thinking. More on this as we go through the book.

[4]   "Attention Span Statistics," *Statistic Brain*, July 2, 2016, accessed March 24, 2017, http://www.statisticbrain.com/attention-span-statistics/.

[5]   Seriously? A *goldfish* can stay on task longer than us? Go figure!

the table chatting about what you thought was an interesting topic.

With that in mind, think about the prompts you're posing to the "miniature you" as he spoons in another bite of your world famous mac-n-cheese. You might want to consider some of these tips for engaging in a healthy Table Talk:

Make it casual.

Formality is the enemy of spontaneity, and we're going for the latter. We want a good discussion, a curiosity-piquing challenge, and an end result of enhanced math fluency. We don't want kids—or adults—to feel forced into a conversation.

To spark a casual conversation, start by saying you "notice" something or "wonder" something, or ask your child if she has questions about a prompt, a picture, or a scenario in front of her.[6] These types of questions naturally frequent our thoughts as we navigate life's challenges, so it's appropriate to raise them aloud with our kids. It may be that you notice something while walking down an aisle at a department store, or wonder something as you're tending to the garden, or stare out the window watching the raindrops make puddles in your front yard. Wherever you are, and whenever these natural, curiosity-filled thoughts come up, share them with your child and let the conversation flow.

Make it meaningful.

Not everything needs to be "fun" to have value. In fact, math tends to gets dressed up in *pseudocontext*—forcing context into a scenario where it doesn't belong in an effort to make it fun. Maybe you've seen the math problem asking you to find the area of the triangle formed on the dog's bandana or the one where you have to work an inordinate amount of mathematical steps to figure out how many apples Sally has. These problems drip with pseudocontext—and your child wants no part of them! Problems like these don't have any meaning for the

---

[6] We will get to more about noticing and wondering in Chapter 9, but essentially we want the child to ask questions in situations where no questions are provided.

students or the teacher, and a good opportunity is lost because they're irrelevant. For example, if you live in Seattle and are talking about crop revenue in Iowa, it will be tough to draw a meaningful connection. Likewise, if you're talking about the costs to repair a home after a flood in Tampa but you live in San Diego—well, you get the point.

Make it authentic.

Being authentic is the best way to set yourself up for a successful discussion about mathematics. Bringing up a task that starts, marinates, and ends in monotony will get old for all parties pretty quickly. I've been the teacher, and the father, who has asked a question void of any authenticity, simply for the sake of posing a problem—and it doesn't end well. I'm not into the discussion, the kids aren't into it, and it ends up being a waste of everyone's time. I've discovered the best prompts for my students, and my own children, are the ones *I* am genuinely curious about, interested in, and want to find a resolution for. Just as valuable, if not more so, is finding out what *they* are interested in and finding something authentic that *they* would grasp onto.

Make it applicable.

As our children get older, their knowledge of their surrounding environment rapidly increases. Social causes become more desirable, injustices become more apparent—and their implications suddenly seem interesting. Whether it is the disproportionate rate of incarceration of different ethnic or socioeconomic groups, the allocation of funds within a city's budget, or how media twists statistics to push a perspective, your child is growing into a world where mathematics can provide clarity.[7] Your conversations may start by evaluating the validity of simple statements such as "four out of five dentists prefer Crest," but when you pay attention, you'll find all sorts gems that connect your child with what matters to her.

---

[7] For the teachers reading this book, I strongly suggest you take a look at Mathalicious.com for some outstanding lessons built on social justice, inequity, and opportunity. If you're a parent reading this, pass that along to your child's teacher; it is a phenomenal resource.

Make it short.

Keep the stopwatch tucked away, cover up the clock on the microwave, and forego the notion that x minutes need to be spent on a specific prompt. As a teacher, I am able to gauge how long a certain prompt will take the class to work through, but that's after teaching the same age students for 180 days for several years. As a parent, I'm much less worried about how long my child considers a question and far more interested to hear what direction it may lead us.

Less is more in terms of finding ways to interject mathematics throughout your interactions at home. So unless there is a sincere desire to go deeper into a concept, keep it fresh and brief. Then when your child asks for more explanation or exploration of a prompt from a week ago, the interaction will feel exciting, not forced.

How do I get started?

Sure, this all sounds great, but I've been looking for something the whole time my child has been able to sit up and haven't landed on a resource (or resources) I love.

Don't worry; my Twitter colleagues have you covered. I joined Twitter in 2012. No, not the Twitter you're thinking of—the one filled with vanity, narcissism, misinformed news events, and ever-tweeting teenagers spouting off about what they ate, who they love, or which survivor should get kicked off the island. My Twitter feed is filled with math teachers who challenge one another to think differently about ways in which we can reach our students. We are building websites and methods to engage students in a variety of ways and doing our best to pique their curiosity—part of those efforts include providing questions to cultivate meaningful education in mathematics.

The resources these teachers have created are shared freely and are currently being used in classrooms around the world. They are also freely available to you as a parent, and they're about as authentic as you can possibly ask for in terms of something you use to start interesting, math-focused conversations with your child. For example, take a look at this Number Talk Image inspired by ntimages.weebly.com:

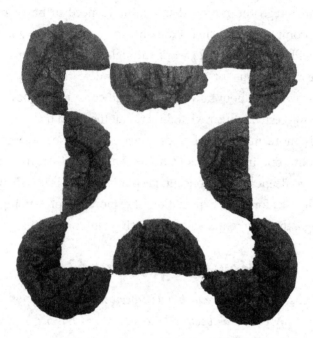

How many cookies do you see in that picture? How do you know?

Take your time as you think about your solution as well as the strategy used to reach your solution. Now might be a good time to show this image to your child, then set down the book and have a conversation.

Ask what your child thinks the answer is and then work through the possibilities. Could there be eight cookies? What about four? Once you've each thrown out a few numbers, ask how or why the numbers might or might not work.

So, what did you come up with?

I got 8... how am I doing?

I got 4... could that be correct?

I got 3... is that possible?

How?

There is more to a solution than the answer itself, and rushing through a prompt does very little to bring value to the conversation as a whole. Take your time pondering the image and let discussion ensue. Don't rush to a conclusion so you can clear the table. Taking your time will bring out more informed, more thoughtful responses.

Related to the topic of speed and online resources, we need to address the issue of timed tests.

As I write this book, my oldest son grows increasingly frustrated with his timed tests at school. He knows the content and feels good about solving each of the problems, but the element of time scatters his thoughts and distracts him from completing with accuracy. During a timed test, all he's focused on is speed. Perhaps your child loves (or loved) timed tests; perhaps you did as a student. The conversation I've had with teachers in early grades about assessments and assignments that include speed as a factor of success only drive a wider wedge between the students who feel successful with arithmetic and those who do not. Over time, the first grader who didn't really like them gets progressively more frustrated and, in turn, becomes less likely to enjoy math.

Dr. Jo Boaler, a professor of mathematics education at Stanford University, has done extensive research and reporting on the area of speed-based mathematics. She has consistently found speed does not equate to success, and memorization is vastly overvalued. In one of her many papers, "Fluency Without Fear: Research Evidence on the Best Ways to Learn Math Facts," she states "the highest achievers in the world are those who focus on big ideas in mathematics and connections between ideas."[8] Please make the time to visit youcubed.org, a website Dr. Boaler and others have created to delve into the idea of math fluency and building strong number sense without the fear of a stopwatch. Although the activities are presented for classroom teachers, they are all applicable to the home environment as well.

Dr. Boaler has published another article on Mathematical Modeling. In it, she details her study findings showing students "who had learned

---

[8] Boaler, Jo, Cath Williams, Amanda Confer, "Fluency without Fear: Research Evidence on the Best Ways to Learn Math Facts," *YouCubed*, January 28, 2015 accessed March 24, 2017, https://www.youcubed.org/fluency-without-fear/.

mathematics through open, group-based projects developed more flexible forms of knowledge that were useful in a range of different situations, including conceptual examination questions and authentic assessments."[9]

While the role of your child's educational system is to facilitate these systems, we can take on a support role at home. The openness Dr. Boaler mentions is achieved through any of the concepts discussed in the following chapters or through others you find elsewhere. Downloading and printing a worksheet to help our child practice his math facts is easy. The bigger challenge—*and* bigger reward—is creating or seizing existing moments to initiate an impromptu activity or conversation with your child centered on math.

Now we've covered some of the ground rules. Let's get to the table.

"The highest achievers in the world are those who focus on big ideas in mathematics and connections between ideas."

– Jo Boaler

[9]   Boaler, Jo. "Mathematical Modelling and New Theories of Learning." *Teaching Mathematics and its Applications* 20, no. 3 (2001): 121-128.

3

Debates

"All right boys, would you rather get the all-day pass to the Harvest Festival or do you just want individual tickets?"

Our mom always gave us perplexing predicaments, and her question at our town's annual fair was no exception. We grew up in a town called Pahrump, a bedroom community to Las Vegas. There wasn't much to do in the one stoplight town, so the annual Harvest Festival was—and still is—an event you simply didn't miss. And we sure didn't want to miss making the most of it! But with so many distractions—friends running past us, the smells of freshly-spun cotton candy and recently delivered rodeo hay bales drifting into our nostrils, and the screams from carnival riders nearly piercing our eardrums—our already tough decision turned into an all-out dilemma. I looked to my nine-year-old brother, two years younger than I was, before questioning my reply as it came out of my mouth.

"Ummm, let's do the individual tickets...?"

Wait! What? What did I just say? I'm at *the Harvest Festival!* So many fun things to do require tickets—the water gun game, the impossible basketball shot, the sledgehammer game—and who can forget the rides?! Clearly, I'd made a bad decision that would *definitely* come back to haunt me. *Idiot! John, you did it again! You missed another opportunity!*

But wait.

Mom and Dad had told us we each had thirty dollars to spend for the day, and we could spend it any way we wanted. The all-day pass was fifteen dollars; each ticket was one dollar, and most rides took one or two tickets. How many rides was I *really* going to ride—especially since my friends were there? If I did individual tickets, I could still go on eight to ten rides, then have more money to get a snack, play the cash-only games, or just save it. *Come to think of it, not too bad a decision after all there, John.*

Even at a young age, I was put into scenarios where I needed to make decisions based on a series of choices. My parents could've handed my brother and me two all-day passes to the fair, and we would have had fun, gone on plenty of rides, and never even wondered what might have been. Conversely, they could've bought us thirty dollars' worth of tickets, placed them into our care, and told us to stay out of trouble until we met up at four o'clock. Similarly, we would have had fun, gone on plenty of rides, and never considered another option. Instead, our parents put us into the driver's seat of the decisions, often with little guidance, and watched us work through what has become a life skill: using math to make better decisions.

How many times have you and a spouse, partner, or friend been headed out for an evening meal, only to volley the dinner decision-making to the other?

"What are you hungry for?"

"I don't care."

"So there's a pizza place, Chinese, burgers, or Thai."

"I don't want to decide; you pick something."

"Would you rather have Italian or Chinese?"

"Gahhhhhh!"

"Soooooo...Italian?"

Would You Rather...?

The indecision around dinner locations is comparable to the indecision around many instances we take for granted. To help kids work through these tough choices, I created *Would You Rather Math*.

Here's an example that you can try with your family tonight:

Would you rather have the bag on the left and share it with one friend or the bag on the right and share it with seven friends?

This seems like such a simple question. In fact, the majority of adults will make a quick decision in their heads—probably like you just did. So, let me ask you: *Which one did you choose?* And, more importantly, *why did you make that choice?*

This question from wouldyourathermath.com is designed to prompt students to do what (hopefully) you're doing right now: thinking about the choice. You're weighing your options. Maybe your choice is based solely on taste. Maybe it's based on the number of friends you want to share with. Or maybe something entirely different drove your decision. Whatever it is, *something* made you choose the bag on the left or the bag on the right. However, the choice you made isn't what I'm interested in, although we'll examine the choice later. I am very interested in *why* you made the choice you did.

In the chip example above, the easy choice to make is the smaller bag. After all, it's less quantity you'll need to share with others, plus it's less hassle to share with one person than with seven. However, upon further reflection, the calculations show you actually get *more* chips per person, as measured by weight, in the bigger bag.

Take a closer look.

Small bag with one friend (1.5 ounces)	Big bag with seven friends (7 ounces)
Total of 2 people eating chips	Total of 8 people eating chips
2 people split 1.5 ounces	8 people split 7 ounces
1.5/2 = .75 ounces per person	7/8 = .875 ounces per person

Present this to a group of students and you'll get two very common mistakes. The first is the deception of the problem itself. They often forget the "sharing with" part and assume one person gets the small bag or seven get the large one. While this is a convenient distracter, I'm more focused on the other stumbling block: the fraction.

Step into any elementary or middle school classroom where the objective is anything related to fractions, and you will see a regular helping of numbers hanging out in the wrong place. In case it's been a little while, here is a visual to jog your memory about a numerator and denominator:

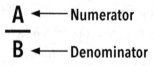

Because we are trying to represent the rate as *ounces of chips per person*, the ounces are set in the numerator and the amounts of people are set in the denominator. This distinction provides an accurate breakdown of chips each person would receive—even though it trips up so many young (and maybe not-so-young) learners.

"*Ohhhhhhhhhh...*"[1]

Having this background knowledge about fractions in your hip pocket, ready to pull into the chat at a moment's notice, would be very helpful. The beauty of the dinner table conversation, though, is one person's struggle becomes a productive family struggle. Look up some information online and you show your child it's okay not to have the right answer, *and* it's *not* okay to move on without trying to find the answer.

Plus, there is more to this problem than ounces per person. The calculations can get a bit more challenging, and this is the junction where some great conversation is bound to occur. When we meet a cognitive roadblock, learning takes place during both the exploration and the decision-making process.

If you let children dig deeper, they may discuss how they don't need to share evenly with all friends, so the bigger bag yields a much higher chip revenue. This makes sense. Now the choice becomes something like this: I can control the distribution of chips much easier with a smaller bag even though I won't get as many chips. Or, "hey, if I take the bigger bag, you didn't say I needed to share the chips *evenly!*"

You can also throw in some variations to the prompt:

- How many friends would you like to share this bag of chips with?
- If you don't like this kind of chip, replace it with a favorite. Does this change your decision?
- What steps did you take in coming to your conclusion?
- Could you share your learning with someone else so they would understand you?

Each of the questions above provokes a different feeling and disrupts the consistent flow of ideas. "Bu ... bu ... but—I just wanted to have my answer!" Yeah, but we're asking you to think a little more about it.

---

[1] Wait, you messed that one up, too? Don't worry; you aren't alone.

Our natural environment frequently poses *Would You Rather* questions:

- Should I take the route my GPS gave me or opt for the back roads I'm familiar with?
- Should I pay cash or use my credit card and get reward points?
- Should I buy cheaper bulk produce or grab the more expensive ready-to-serve bag?

Each time we make these decisions, we take into account a variety of factors. Often, price or time is the deciding factor. Every now and then, we have to consider weight, calories, height, efficiency, speed, distance, and especially, personal preference. Regardless, we inevitably lean toward one option, and our choice will very likely have something to do with mathematics.

Let's take another example: traffic.

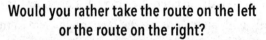

Would you rather take the route on the left or the route on the right?

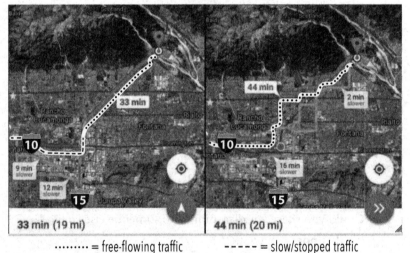

33 min (19 mi)   44 min (20 mi)

·········· = free-flowing traffic   ----- = slow/stopped traffic

At first glance, this is a pretty easy decision. Even though you'll be sitting in a sea of red tail lights, taking the route on the left saves you eleven minutes. However, upon further inspection, there's more to it than simply comparing the minutes of the drive. For twelve miles, you'll essentially be stuck on a freeway-shaped parking lot. The accident right after the point where I-10 merges onto I-15 backs up the entire flow of traffic by approximately nineteen miles (not seen on the screen).[2]

Once you get past the accident, it's smooth sailing, but *getting there* is the issue. Are you really going to sit in traffic, gripping the wheel with a pair of white-knuckled fists after a long day of work? Also you're traveling around 4:30 p.m. on a weekday, which means anything could happen to the traffic between now and the time you get to open road, possibly delaying you even more.

The other option is to take side streets all the way home.[3] You'll gamble getting the green lights, making the correct turns, and dealing with other drivers stopping at their red lights.[4] Taking this route also means you're willing to travel more cumulative miles. You will be restricted to slower traffic zones and might go through a school zone right around the time school gets out. However, you will be moving at a rate compatible with your surroundings. (Generally, when we're on the freeway, we expect to go fast. When we're on surface streets, we can hope to hit all the green lights to go faster, but we don't expect to.)

So … Which route would you take? More importantly, why?

Once again, we adults are accustomed to making these kinds of decisions. Some of us trust our navigation apps 100 percent of the time. Others consider their app directions, but make snap judgments based on experience. But our children haven't had those experiences yet, so mathematics can help them figure out which route would be

---

[2] The worst part of this whole scenario? It is commonplace in Southern California. And actually, we're the lucky ones. If you travel closer to Los Angeles, you might as well forget about driving. Seriously.

[3] For now, we're going to ignore the third option which would be to take side streets until you get north of the accident and then jump onto the freeway. Too many screenshots I wasn't willing to risk taking while sitting in traffic... this time.

[4] Not that Southern California should be considered rural, but there are statistics out for rural and urban driving accidents. Spoiler alert: rural areas are more dangerous. npr.org/2009/11/29/120716625/the-deadliest-roads-are-rural

better. In the image above, the total distance of the route on the left is approximately 15 miles; the route on the right is approximately 20 miles. Does that change your decision? If so, how? If not, why not?

For what it's worth, I chose the option on the left. I'm a bit of a glutton for traffic punishment, and I opted to save the eleven minutes—exchanging more time staring out my windshield and listening to my favorite podcast for making far more turns in an unfamiliar area.

Here's another, with a little bit less urgency:

Would you rather travel along the path of the triangles or the circle's circumference?

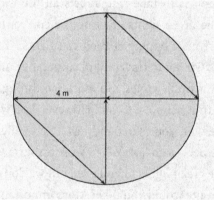

If you opt for the triangles, it would be helpful to know that the radius of a circle is the same no matter how you draw it. As long as the segment starts in the middle of the circle and goes straight to an edge, the radius in this circle will always be four meters. Knowing that and assuming that we are looking at right triangles, here's one way to approach the problem:

4 meters for the first side + 4 meters for the second side = 8 meters

For the third side, you'll need the Pythagorean Theorem, $a^2 + b^2 = c^2$ where a = 4 and b = 4. When you do this, $4^2 = 16$ so $c^2 = 32$. Taking the square root of 32 gets you 5.66, so all three sides will add up to 13.66 meters. Do this twice and your distance traveled is 27.32 meters.

Whew . . . too intense?

Especially if it's at home, estimation could help. The long side connecting each radius looks a little longer, so let's call it 5.5 meters.

4 + 4 + 5.5 = 13.5 and since there are two triangles, doubling this gives us 27 meters.

For the circumference, a formula might help, which is C=2πr, where C is the circumference, π (pi) is typically understood as 3.14..., and r is the radius. Plugging the values into the formula looks like this:

$$C=2\pi r$$
$$C=2\pi(4)$$
$$C=8\pi$$
$$\textbf{C=25.12 meters}$$

Yes, the circumference is a shorter path, but which one would you rather take? Why? That's where the fun begins.

And what about this one:

Would you rather mine 1 kilometer of Earth for 2,335 grams of gold or mine through the entire crust for 500 kilograms of gold?
*Whichever option you choose, you must also pay for
the cost of mining the gold.*

Photo Credit: Chuck Stevens (my dad) with full permission and rights.

Think about the types of questions that might come up before we answer the prompt:

What is the value of gold these days?

How much does it cost to mine gold out of the ground?

What am I going to do with all that gold?!

How thick is Earth's crust?

How long would it take to mine that deep?

Has anyone ever thought of doing this before?

How hot is it down there, anyway?

Is this even possible?

Basic online searches will return answers to the majority of your questions, but the prompt remains: which one would you rather choose, and why? If you are talking through this with an elementary schooler, you may keep the numbers whole and just go after the amount of gold; would it be worthwhile? If you are talking through this with a high schooler, maybe you're going to get into the deeper question about feasibility. Maybe you'll plan out the costs as you get deeper into the Earth. Or maybe, just maybe, you keep the numbers whole and just go after the amount of gold.

Bringing It to the Table

Pose a question your child would be curious about. If you don't live near farmland, asking if they'd rather get a certain yield from crop A versus crop B is a lost conversation. If she isn't into sports, asking which basketball player she would rather have on her team (based on statistics, not names) would mean nothing. If she isn't into lifting heavy things, asking about carrying a certain bucket versus another is, well—empty. Instead, think about the decisions you make on a regular basis and how those might apply to a conversation with your child—purchases to make, routes to take, choices to ponder. Look around. There is a probably a *Would You Rather Math* prompt that you can bring to the dinner table.

I created the website as a way of collecting as many of the naturally occurring struggles as possible. Whether it was a driving route, bulk versus individual, credit card versus cash, or lease versus ownership, we encounter a lot of situations where math plays a role in how we make decisions. The best part about creating the site, though, has been the guest contributions from people all around the world. People from other parts of the globe have submitted all kinds of questions like which phone plan is best, or if it's better to carry a heavier object a shorter distance than a lighter set of objects a further distance, or whether it's better to take kids trick-or-treating in a high-density neighborhood with little treats or a low-density neighborhood with big treats.

While none of the prompts on the site are aimed at solving the world's problems, the intent of the website was to put something in front of kids that would encourage argument and disagreement with a safe topic. We aren't discussing politics, economic policy, or who should have the toy after dinner time; rather, we are gathering to work through a not-so-hypothetical scenario.

One of my favorite classroom discussions that arose from a Would You Rather Math prompt was this one:

Would you rather have a stack of quarters from the floor to the top of your head OR $225 cash?

What was supposed to take seven minutes of class wound up taking seventeen, and it was worth every second. My students debated what you would do with all those quarters, where you would turn them in, the percentage that the coin machine would take from you, and many other interesting ideas, all tied to that one simple question.

I bring this up because, among other reasons, putting options in front of your child and having her make a decision—then justify it—can yield some pretty amazing insights.

Encourage

- Disagreement
- Doing research
- Having more than one solution
- Justifying the decision
- Finding various examples around the house or elsewhere

Avoid

- Fixating on one correct solution
- Providing too much help
- Doing the research for the child
- Focusing on the solution rather than the process
- Using a prompt with little relevance

For more prompts like these, head to wouldyourathermath.com and explore. When you come up with your own, starting a great conversation, email me and I'll be more than happy to add it to the list!

4

Estimations

Like many Christian-American families with little ones in the house, we celebrate Easter. Nothing exorbitant, but we put up decorations, dye some eggs, and spend time together as a family. Oh—and hunt for Easter eggs—an absolute must every year since I was a child.[1] While setting up for Easter is joyous, putting away the décor...not so much. One particular year, Easter clean-up was beyond aggravating. Our boys ran around the house throwing Easter basket grass at each other. When the grass throwing didn't incite enough reaction, they grabbed the empty plastic eggs and rolled them down the hallway. Amidst fits of laughter, the mess grew even bigger. It was driving me bonkers. When I had seen (and heard) enough, I spoke up in my sternest dad voice.

"Stop it! We need to clean up!"

(No response.)

"I said... you're gonna break something again!"

(More laughter from down the hallway.)

"BOYS, GET OVER HERE RIGHT **NOW**—and bring the eggs!"

What started as a boil of fatherly frustration as my kids ignored my requests for civility suddenly turned into a simmering opportunity. Keeping my composure and rapidly switching gears from "Mad" Dad to "Math" Dad, I brought the now-inquisitive little ones to a large glass vase and bulldozed the smattering of empty unbroken plastic eggs together. Both boys were sitting criss-cross-applesauce and looking up

---

[1] True story. My parents hid Easter eggs for my brother and me when we were in high school!

at me, knowing my change in demeanor meant they were no longer in trouble and were about to be given a job.

"Okay. So here's the deal: I need to know how many of these eggs (holding one of the empty plastic eggs) will fit into this jar. Can you figure it out—without stuffing the vase full of eggs? And boys, keep your hands to yourself."

No response to my question. Nolan simply put his serving of hollow plastic eggs into a big circle and declared, "FIFTYYYYY!!!" because fifty was the biggest number in his three-year-old universe and the quantity of eggs in front of him far exceeded his counting limit of nine. Sure thing, kid, stick with fifty.

Luke, five at the time and very determined, started building layers. He knew he couldn't put eggs *into* the vase, but nobody said laying them *outside* the vase was forbidden. I could see his mind at work as he imagined how many tiers of eggs would fit inside the glass container. After nearly ten minutes of heavy thinking and a whole lot of counting and recounting, he looked up out of the corner of his eye and, with a sheepish grin, declared, "Thirty–two! Am I right, Daddy?"

Like I'd done with my students for years, I replied with the least amount of confidence possible, "I dunno. Let's find out together."[2]

We then proceeded to fill the vase—one egg at a time, counting each one as it plummeted to the bottom of the cylinder and made a hollow clunk—until it was nearly full. "Forty-four, forty-five, forty-six, *forty-seven!!!*"

Crushed by being wrong, Luke slouched, puffed out his bottom lip, and waved the emotional white flag. Even without years of a formal education, he knew his guess was "wrong" and he really wanted to be right. Nolan, on the other hand, was still sitting in his attentive legs-crossed position, waiting to hear the end of the story. After all, being wrong didn't really matter when all he'd done was blurt out the biggest number he knew!

---

[2] Admitting a gap in knowledge—even if it doesn't actually exist—creates a great opportunity. We parents and teachers are expected to have all the answers, so it's nice to share the times we don't.

"Let's see if we can help Nolan out a little," I proposed to Luke. "Can we fit more eggs on the pile?"

"Yusss, fine," my disgruntled five-year-old murmured through his still-pouty lip.

Adding one after another, we kept counting: "Forty-eight, forty-nine …." When we added the Nemo egg and announced "Fifty," Nolan, the wily little runt, broke from his attentive posture, jumped up, and over-excitedly screamed out in victory, "Mommy, Mommy, Mommy, *Mommy!!!* I gawt it wight!!!"

Estimate This

Something about the task of estimation is spectacular. In it, all *rightness* and *wrongness* are gone and, in their absence, comes the freedom to take a risk. Watching kids at play makes it clear that humans are designed to estimate first—then act. On the playground, kids swing back and forth to see if they can actually make it to the next monkey bar. We use our fingers and count imaginary steps in the air to determine if an object made with certain materials will be tall enough. We even flex our knees, squint our eyes, and clinch our hands on the steering wheel in anticipation of hearing an unnatural scraping sound as we watch to see if our estimation of the space between the side mirror and the garage door was right.

Estimation is a natural occurrence—easy to foster in our kids and especially at the dinner table—but because it is so automatic for adults, it's easy to miss simple opportunities to get kids involved in estimating. Fortunately, estimation180.com is a website stacked with over 200 *free* prompts that will get the curiosity flowing. The creator of the site, Andrew Stadel, uses his life experiences and surroundings to share estimation opportunities. Here are a few more estimation ideas from estimation180.com to try with your family:

How many sheets of toilet paper are on this roll?

Estimation180.com

Have your kids grab a fresh roll of toilet paper from the bathroom and put it in the center of the table.[3] Then, ask them, "How many sheets of toilet paper are on this roll?" and let the estimations begin. The question is quite simple, but the responses elicited are far more complex.

Kick off the conversation with "What number of sheets is too low? Only _____ sheets on this roll are impossible."

Depending on how many kids are around the table and how enthusiastic they are, collecting their responses may take a little while. Once you've got them, head for the upper bound:

"What number of sheets is too high? It is impossible to have _____ sheets on this roll."

Once again, take responses from all at the table. You're not looking for precision here; in fact, you don't want it. What you do want is for each person around the table to think about what a risky estimate would be for the lower bound (too few) and upper bound (too many). This thought process confines your estimation to a certain range, making it more likely you'll land on a number *close to*—but not *exactly*—the correct one.

Once you have your bounds set, it's time to encourage the risk-taking: "All right, we know it can't possibly be _____ (lower bound) sheets, and there's *nooooo way* it's _____ (upper bound) sheets. So what's your guess?"

With each kid laying out his guess for the group, everyone has a fair chance of getting the correct answer (unless, of course, you're the one buying the toilet paper and paying attention to the packaging information!) The reveal, although anticlimactic for those who didn't guess the right number, gives way to sighs of relief as kids realize they weren't the only ones with the "wrong" estimation.

---

[3]   A fresh roll of toilet paper is better than the fake plant or handful of fresh flowers blocking everyone's view anyway. Plus you now have something to talk about at the dinner table!

Estimation180.com

While this prompt is enough in itself to host a strong conversation, don't stop there. If your youngster is in elementary school, ask her to find the difference between her response and the correct solution. Was she off by one hundred? Fifty? One thousand? Learning she was off by just as much as others at the table is a lot more encouraging than simply discovering she got the incorrect answer.

If your youngsters aren't so young anymore—maybe middle school or high school aged—take it a step further. Ask them to find their percent of error using this formula:

% error = (student answer - solution) / solution

For example, my guess was 600 sheets. The correct percent error is found by:

$$\% \text{ error} = (600 - 425) / 425$$
$$\% \text{ error} = 175 / 425$$
$$\% \text{ error} = 34.1\%$$

This means I was 34.1 percent *over* the correct solution. If the percent error is negative, I would've been 34.1 percent *under* the correct solution.[4]

Now, don't be alarmed when you're off by thirty, forty, or one hundred percent—or even more. You've taken a risk and responded on a previously meaningless topic. Your child may also need a bit of encouragement to move on—especially if the feeling of not being "right" is new. Considering you've read this chapter, if you do the shopping for toilet paper, it might be a good idea to be intentionally wrong.[5] But in this scenario, you're locked in. You saw the "425 sheets" on the packaging, it was in your sights and you used the number to your favor. Fortunately, there is a sequel for tomorrow night's dinner table:

How many sheets are on the smaller roll?

Estimation180.com

Employing the same questioning techniques, the second day of this Estimation 180 series builds on prior knowledge. Last night, you threw around wild guesses based on nothing. Tonight, you have a frame of

---

4   Super validating! And the happiest I've ever been about being over 34 percent from the right answer.

5   I often employ this strategy. While we will talk about this later, there is a lot of value in not having all the right answers all the time.

reference and you're ready. The roll on the left is the roll from the previous image. Tonight you want to know how many sheets are on the smaller roll.

Once again, asking for the lower and upper bounds is important.

How many sheets is too many?

425

Why?

And there it is! This simple question—*Why?*—is the reason estimation tasks are so important. Your child now knows she is being held accountable for her responses and will be asked to explain herself.

Her answer might sound something like, "Well, because the roll on the left is way bigger, and it's the one with 425 sheets we saw last night. So, the smaller roll has to have lots less sheets."

Bingo! Her response might be a small victory, but it's a victory nonetheless. With this in mind, you can move forward to find an upper, a lower, and an estimate. As before, the discussion is more important than being correct.

After estimates have been fielded, it's time to count. Yes, Andrew tore sheets of toilet paper off the roll as he methodically counted each one.[6] Imagine the suspense around the dinner table as you collectively count expended sheets of toilet paper!

Just like my son's agonizing "defeat" of his erroneous egg counting, there's a good chance everyone will be incorrect about the remaining number of sheets. But the value is in the conversation about how each person got to his number and what he needs to adjust for next time. Did you take your original guess and double it? Maybe you looked at the diameters and compared the large roll to the small roll, deciding that it was one-third the size. Whatever it was, your new number was based on your old number and explaining how you got there is worth the time it takes.

---

[6] For the video reveal of this prompt, go to vimeo.com/51322959 and enjoy Andrew's work. It's a blast to watch.

Here are some additional prompts from the website for you to ponder:

How much do the dumbbells on the left weigh?

Some questions that may go into the thought process:

Estimation180.com

- What else do we have to compare the purple dumbbell to?
- How much bigger is it than the one on the right?
- What do dumbbells usually weigh when they are smaller like that?

The answer is on the website, but the value is in the guesses.

How many gobble napkins come in the package?

Estimation180.com

Some questions that may go into the thought process:
- What size packages do napkins usually come in?
- How thick is one napkin?
- There are plates underneath; could those help inform my decision?

Bringing It to the Table

Few people have made me realize that even the youngest kids can benefit from experiencing high-level mathematics like Andrew Stadel has done. Whether it is his website, a presentation, or simply a conversation, Andrew is sure to share something that will make all of us think.

From Andrew Stadel, founder of Estimation180.com

During the early days of using *Estimation 180* challenges with my students, I found their accuracy and success was the result of the connection between their strategic thinking and prior knowledge. For example, on Day 1 of school, students estimate and discover my height. The students then spend the remainder of the week estimating the height of my wife and two children, whom they have never met. Using my height for reference helps them figure out how tall my family members are.

There's great value in classroom discourse and in how students get invested in both their estimates and reasoning. In my class, student conversations have gotten so heated that, at times, I've felt as if I were a moderator at a political debate. In reality, students were so passionately explaining their thinking behind the number of red vines in my hand or would get upset that I didn't squeeze just one more cheeseball on a tray because it would match their estimate. I'll never forget reviewing inequalities with my students midyear, and I brought back an almond estimation challenge from the beginning of the year. I put the visual of a quarter cup of almonds on the board, and immediately, a large number of students remembered the exact number in the jar, even though they couldn't remember the two questions I assigned for homework the night before. There's something really sticky about the visuals and conversations that come from *Estimation 180*.

As a parent, I look to provide my children with that stickiness of estimation. With a six-and-a-half-year-old son and four-year-old daughter, I have plenty of opportunities to estimate with my children. For example, my wife ordered four boxes of smelly markers through the mail because she loves giving the markers as birthday gifts to other children. When the shipping package came to our house, my children did not know how many boxes of smelly markers my wife had ordered. However, they knew the size of one box of markers because we already own a box of smelly markers. I looked at the size of the package in which the markers were shipped and asked my children, "Before we open it, how many boxes of markers do you think came in the package?" It was fascinating to see my son break apart the package in his head while drawing in the air the numbers of marker boxes he was visualizing. He estimated four boxes of markers. My daughter simply said, "100!" because she's silly like that and genuinely wished there were 100 boxes of smelly markers because she loves smelly markers. I would classify her 100 boxes as an answer that is too high. It was amazing to see my son get excited after opening the box and discovering there were four boxes of smelly markers inside. His thinking and plan worked! Even if he was off by one or two boxes, we would have celebrated his thinking and plan. I reflect on the fact that a simple question like, "How many boxes of smelly markers are in the package?" was so powerful because it gifted my children with the opportunity to think and estimate, using numbers to make sense of the world around them.

Encourage

- Ask questions. Oftentimes, more information improves an estimate. What questions could you and your children ask so the answers would help make a better estimate?
- Think time. Encourage your child to analyze a visual and look for clues.
- Reflect. Use information from previous days or prior experience and apply it to the current context.
- When your child gives their estimate, encourage them to follow it up with reasoning.
- Give a chance for revision. Pause a video answer somewhere near the middle and offer your child the opportunity to revise their estimate now that they have more information.
- Think of an estimate as a rough draft and look for opportunities to refine your estimate from either hearing or seeing new information.

Avoid

- Avoid scorning children if their "too low" is something like one or zero, because they're actually correct with such a safe answer. Instead, encourage them to, as 3-Act framework creator Dan Meyer says, "be braver" next time.
- Avoid accepting "I don't know" or "I just guessed" from children when asking them for their reasoning. Encourage them to take an extra minute and reflect on the information in the image or the prior knowledge they used to make their estimate.
- Avoid rushing to reveal the answer as you could potentially rob your child and yourself of valuable questions, think time, and conversations. Feel free to walk away and revisit the challenge after a few minutes.
- Avoid focusing on the answer and how close or far away you were. Celebrate the thinking.

I encourage you to visit Estimation180.com for good prompts and quick conversation starters that will interest everyone at the table. From cheese balls to spare change, he has an image for just about any estimation you could make around the house. But don't stop there. Many estimations happen naturally in our regular routine; be on the lookout for them and present them to your child.

Whether you are working in the garage, kitchen, or craft room, or taking an afternoon walk, you can find opportunities to initiate a conversation with your child about estimation and its relevance to improving math fluency. Capitalize on those moments and enjoy the conversations—and learning—that ensues.

5

Patterns

M y boys and I were in the process of cleaning up their toys before bedtime—a daily routine filled with more frustration than fun. The boys had been playing with Play-Doh, a product second only to glitter on my list of "most despised" crafting supplies.[1] At first, I put the Play-Doh containers on the kitchen counter because they were distracting the boys from cleaning up. They would grab the containers, stack them up like they were a tower, then knock them down, laughing and deviating even further from the goal of a clean house. Something had to be done, so I collected all the containers and put them out of reach. However, while sitting atop the kitchen counter, those jars of mush sparked an idea.

"Oh hey, boys! Check this out. What does this make?" I asked.

"*A triangle, of course.*" My six-year-old was irritated that I interrupted his cleanup to involve him in yet another experiment. He's a smart one—smart enough to know when he's being pulled into

---

[1] We actually have a "no glitter" rule in our house. And it has nothing to do with masculinity or the fact we are raising two boys; it has *everything* to do with the mess it causes! You can't get it out of your hair—or the carpet—and you can't ever completely clean it up. Play-Doh is *almost* as bad.

one of my games. Normally, he enjoys these scenarios, but redirecting him to clean up hadn't set the right mood.

"Hmmm. How many jars did it take to make the triangle?" I asked.

"Twooo. No, wait. Freeeeee," my four-year-old beamed as he answered. He knew something else was brewing and he was on the right track.

"Oh yeah? Well what about this?"

"Still a triangle. DUHHHHH," Luke sighed. Still unamused, but now eagerly looking for a challenge, he began to doubt anything would be complex enough to make him curious. Clearly, he thought this must be a task better suited for his little brother.

"Ok, fine. How many more containers do I need to make another triangle?"

"Bewilderment" is the only way to articulate the level of confusion on their faces. They were befuddled—as if I had taken away something I wasn't supposed to touch. The boys seemed to feel tricked; at the same time, they were fascinated and wanted to know the answer. But now, I was in a tough spot. I didn't have any more jars of Play-Doh so we had to use some creativity to answer the question.

Both boys accepted my challenge, but went in different directions. Luke, cerebral and determined, went to his notebook and drew enough circles to represent the pattern and its missing information. Scribbled onto his page is a flurry of his thoughts, a little bit of frustration, and his success.

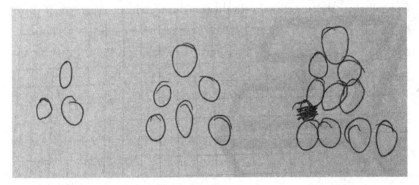

Nolan took a much different approach. Since he knew he couldn't touch the existing jars, he grabbed a stack of LEGO bricks from the toy bin for a bit of tactile assistance. As you can see below, Nolan built the missing portion of the triangle then counted how many pieces he used to make the model.

Two different approaches, two different age and ability levels—but two correct solutions to the challenge. If the boys had been older, we would have started developing the pattern for the next step and the next one, and so on. We would've talked about how reasoning allows us to extrapolate a set of data in a way algebra supports. But in this instance, we celebrated the fact both boys created the missing pieces of a triangle in two completely different ways. This success is a big reason to fall in love with visual patterns.

Visual patterns, a concept formalized by Fawn Nguyen, are the idea of extrapolation—or thinking about how something smaller can grow bigger—and the steps it takes to do so. In the example above, my boys worked (mostly) together to build the third step in a pattern. If they had been older and more jars of Play-Doh were available, they would've continued with successive steps until they realized there must be an easier way to determine the *nth* iteration of a pattern.

At which point, we would've discussed the algorithm for this pattern:[2]

$$t_n = \frac{(n + 1)^2 + (n + 1)}{2}$$

For each step in the pattern, plug the number of the step into the formula. For example, for step 3, you would get:

$$t_3 = \frac{(3 + 1)^2 + (3 + 1)}{2}$$

$$t_3 = \frac{(4)^2 + (4)}{2}$$

$$t_3 = \frac{16 + 4}{2}$$

---

[2] Keep in mind that this would be for *older* kids, not your second grader who is reluctant to sit down and ponder the mathematical wonders of the world.

$$t_3 = \frac{20}{2}$$

$$t_3 = 10$$

Ten jars are needed for step 3.

For step 11, plug the eleven into the formula and you should get:

$$t_{11} = \frac{(11 + 1)^2 + (11 + 1)}{2}$$

$$t_{11} = \frac{(12)^2 + (12)}{2}$$

$$t_{11} = \frac{144 + 12}{2}$$

$$t_{11} = 78$$

Seventy-eight jars are needed for step 11. For step 50, you'll need 1,326 jars, and so on. The sooner kids see a pattern, the better.

One scaffold I built in for them was the inclusion of *some* pieces to the next step, rather than asking them to come up with something on their own. Had they been older, I likely wouldn't have done this, but for their ages, the last thing I wanted to do was create a prompt so challenging they'd shut down. In this case, it was just enough to make them think and be willing to work on an idea.

Recognizable patterns can be found anywhere—from the railroad tracks to the airport to the flooring in our homes. The ability to project out—or extrapolate—a pattern is a helpful skill as children begin learning about linear modeling and variables. If this makes your skin crawl, think about Algebra 1. This course in a student's mathematical life focuses on moving from the concrete to the abstract. Consider this example from visualpatterns.org:

When I show this to a group of adults, their mental wheels immediately start turning as they consider what the fourth and fifth steps look like. Then, I ask them to consider a tougher question: What does the forty-third step look like? You can choose a step number depending on the age of your child at home. For my little ones, I'm going to pick the fifth step; for my middle-school students, I might choose the fifteenth. With anyone older, though, I'm diving right into the forty-third.[3]

Those in the younger crowd, may need to draw out each step, or like my son did, find some sort of tactile tool around the house that can be used to support their thought process. You might want to prompt your child by asking questions like, what was added from the first to second, second to third, and so on? In order to keep the pattern going, what do you need to keep doing?

As you and your child work through these examples, don't worry about how long it takes. Time is *not* of the essence. Remember the attention span we spoke of in the beginning of the book? It doesn't apply here because the prompt itself has built a level of curiosity that removes attention span limitations, even if only temporarily. On top of that, we know full well no youngster is going to figure out the

---

[3] You know your child better than I do. Choose a step barely out of her reach which still allows you to hold a valuable conversation with her. You've got this.

algorithm,[4] or the process used to solve a problem, in seven seconds. So, be patient. There is real value in letting your child struggle through a few steps, building up his capacity to show step five (or whatever step you put in front of him).

Be patient. There is real value in letting your child struggle through a few steps.

For the pre-teens and tweens, you'll want to find the balance between challenge and burden. Even as an educator, I have to find this balance, and all too often, I misjudge it. There is no formula or sure bet for the "just right" difficulty of a problem. For patterning problems, err on the side of challenging and be ready to support your child if (or when) the struggle becomes too much. In the example above, it will be a grind to write out each iteration leading up to the tenth step, but it's possible. If your child chooses this option, showing her how each step increases by a certain amount every step might be the lightbulb she needs to make the connection between the drawn-out pattern and the formula (or algorithm) to make future calculations easier.

Algorithms were designed to make life easier for us. They allow mathematicians and scientists to wrap their heads around changing criteria. When we were in elementary school, the abstract idea of consistent change didn't apply to the majority of our math work: add these numbers, divide these numbers, skip count by this many, etc. Once we got into middle school, the concept of mathematics grew larger than a simple calculation and demanded an algorithm. Visual patterns strongly encourage learning the algorithm because counting each step is so difficult (or tedious); students start to understand that it is actually *easier* to learn the formula.

---

[4]   An algorithm is a term that mathematicians and scientists use to define a process or set of rules to solve a problem. I could have used the word "formula," but it is more than that. Algorithms help us streamline the way we process information, whether we think about it that way or not.

We want our middle and high school-aged children to get to that place of understanding as quickly as possible. The first three steps establish the pattern, but the fun images end there. What does it look like when they get to step seven? What about step fifteen? Finally, what does the pattern look like in step forty-three? Drawing step forty-three of almost any pattern will be cumbersome, and people will lose interest in a hurry. Finding a better, more efficient way to represent the model is in everyone's best interest.

Going back to my son and the containers of Play-Doh, he was capable of building step four. If he were ten years older and experienced with creating formulas, I would've gladly asked him to "build" step forty-three, confident that he would opt for the algorithm over reaching for more containers.

Formulas make patterns easier to understand. Think about how extrapolation and algorithms affect other areas of education. We use them to find an easier way to:

- count the distance to stars (light years)
- measure how small a molecule is (scientific notation)
- write down each step of a savings account (compound interest)
- write each year of radiation decay (exponential growth/decay)

The need for formulas becomes evident as we get older. Using them with your child to solve seemingly simpler tasks will demonstrate how he can use math to figure out more challenging assignments in the future.

> *The essence of mathematics is not to make simple things complicated, but to make complicated things simple.*
>
> —S. Gudder

Just imagine! Math as a way to *simplify* the world around us. Yeah, I dig it!

Here are a couple of more patterns to work out at your family's table.

How many LEGO bricks will there be in Step 43?

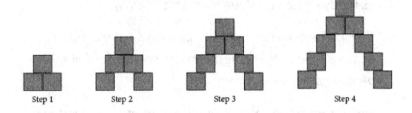

How many dots will there be in Step 43?

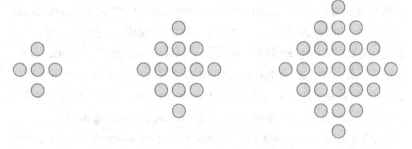

Some questions to help inform your decision:
- How many dots are in Step 1? Step 2? Step 3?
- How much is the total increasing each time?
- Is it going up a consistent amount?
- If not, what else might be happening?
- What would Step 4 look like?
- What would Step 5 look like?
- How about Step 10?

Bringing It to the Table

It's easy to find patterns to play with around the home—and it can be fun! But don't take my word for it. Take a look at what Fawn Nguyen says about what visual patterns can do for your child's cognitive development and its effect on the learning process.

From Fawn Nguyen, creator of VisualPatterns.org

G. H. Hardy wrote in *A Mathematician's Apology*,

> *"A mathematician, like a painter or a poet, is a maker of patterns. If his patterns are more permanent than theirs, it is because they are made with ideas."* [5]

I think it's wonderful that one characterization of mathematics is the study of patterns because patterns are all around us! One of my favorite patterns is the Fibonacci sequence that occurs in nature. I've always loved math growing up; something is reassuring about the order and structure of it. Undoubtedly, my father, a math teacher, cultivated and nurtured that love. I would spend hours and days working on a math problem or puzzle, and I'm sure growing up poor in Vietnam, without toys to play with and trips to go on, doing puzzles occupied my spare time fully without costing anything.

My first encounter with visual patterns came late in college when I took a series of courses called "Math and the Mind's Eye" at the Math Learning Center in Portland, Oregon. Up until that point, I was always given equations to solve. But through visual patterns, I was asked to *create* equations. That left a powerful impression on me! I began gathering patterns and creating my own. Then in 2013, I built visualpatterns.org and invited others to contribute to the site. We now have over 200 patterns.

Back in Chapter 2, John speaks of encouraging your child to "notice" and "wonder" about their surroundings, and that's exactly what I would encourage you to do with a visual pattern. There is no right way to notice or wonder about something. What do you notice about the pattern? How is it growing (or shrinking)? Is there

---

5   Hardy, G. H., *A Mathematician's Apology*. Alberta, Canada: University of Alberta Mathematical Sciences Society, 1940. https://www.math.ualberta.ca/mss/misc/A%20 Mathematician%27s%20Apology.pdf.

any part of the pattern that stays constant? What do you wonder about the 24th step, or the 56th step, or *any* step?

I introduce each visual pattern by asking students to draw what they think the next step might look like. I then ask for a "quick" or rough sketch of some step farther down the series, such as step 50, to check for understanding. We try to describe what any step may look like—given step *n*, how many objects are in that step? When we are comfortable with this description, we move on to represent this pattern with an equation, changing our verbal description into an algebraic one.

I gave my sixth graders this problem from visualpatterns.org, and in Google Draw, I asked them to draw the next step, then draw the 50th step, and write the equation for this pattern.

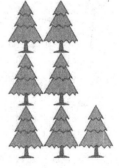

Here are the sketches from three students.

Student 1:

Step 4

50

50

Equation

Step # x 2 + 1 = any step

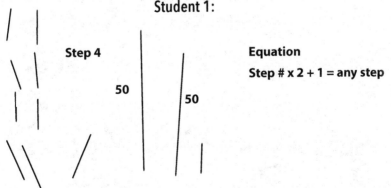

Student 2:

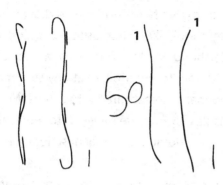

Equation: 2n + 1

Student 3:

$$T = 2n + 1$$

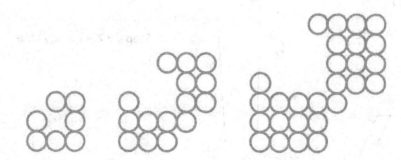

Pattern #5 from visualpatterns.org is one my favorite patterns because of the various ways one can see this pattern.

Four of my eighth graders share how they see this pattern. They have been working with visual patterns since 6th grade, so we skip drawing the next step (or drawing any step) to save time and move straight to working on the equation for the pattern.

Student 1: I see a column of $(n + 1)$, and another one up here. Then I see two squares, each side is always (n). Plus this extra one tucked in here. So my equation is $C = 2(n + 1) + 2n^2 + 1$.

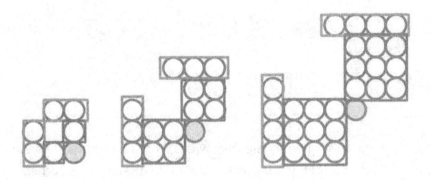

Student 2: So I see on the left $(n + 1)$. Then I see a square of n by n. Then I see a rectangle of n by $(n + 1)$. And these two leftovers. My equation is $C = n + 1 + n^2 + n(n + 1) + 2$.

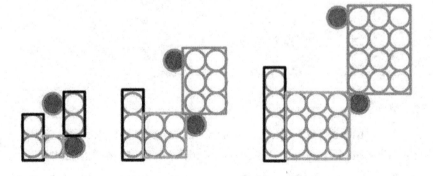

Student 3: I see two rectangles of the same size. Each one is n by (n + 1). And then the three circles here. My equation is $C = 2[n(n+1)] + 3$.

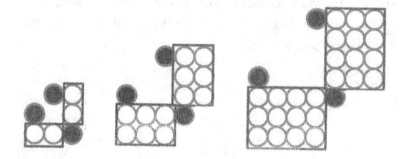

Student 4: I see two squares that overlap. Each side of the square is (n + 1) long. Then there's always one circle tucked in the corner. Because the squares overlap, I have to subtract the overlaps to not double count. My equation is $C = 2(n + 1)^2 + 1 - 2n$.

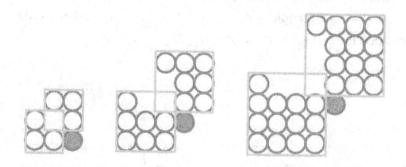

Since these are *visual* patterns, it's important that students describe how they *see* the pattern growing and not just tell you how many objects there are in a particular step. For example, a student submitted pattern #217 below to visualpatterns.org, and the structure of her pattern would be lost if we only summarized her pattern as the number of circles is twice the step number. Sure, this is the correct answer, and the goal is to come up with the equation to develop algebraic reasoning, but the *beauty* is also in how the pattern presents itself.

Encourage
- Persistence on a problem
- Drawing out the steps as a beginning support
- Finding a formula, depending on the child's age

Avoid
- Providing the formula for them (if you know it)
- Relying on the images
- Choosing something *too* challenging

6

Organization

When I was a kid, *Highlights* magazine showed up in our mailbox every month. *Highlights* had a lot of rich content for kids—from news articles and stories to puzzles and challenges. I was always the need-to-answer-all-the-questions kid, while my brother was the tag-along-watch-and-wreak-havoc–on–my-flow kid. So every time we received a new issue, we fought to see who got the first crack at the fun stuff.

My favorite section was always the puzzles, and I loved figuring out the solutions as quickly as possible. One in particular was the "What's Wrong?" puzzle. On one page, there were two images that were very similar but not the same. For example, there may be a carnival scene with a roller coaster in the background, a girl holding balloons and an ice cream cone, and a farmer walking a donkey through a dusty path into a corral. Looking through the two similar scenes, the reader's job was to find out what is different between the images. The bow in the girl's hair that doesn't belong, the extra tooth on the donkey, and the balloon with the string defying gravity were all clues to circle and find a solution. Thinking back, those puzzles sure were validating. Right was right. Wrong was wrong. Life was easy.

Ahhhh, the good old days—when validation came as easily as a calculation. Today, students are asked to formulate hypotheses, come into a discussion with justifications, and reason with a problem from multiple angles. They are asked to solve open-ended and ambiguous problems that require more thought and understanding than the simple

"What's Wrong?" puzzle ever did. Yes, math class can be more challenging than it was when we were in school. However, it can also cultivate individual thought processes and new ideas and serve as a launching pad for those *oh-yeah-I-get-it* moments that a right or wrong answer couldn't offer. As parents, we can either think of this shift as intense and overwhelming or inspiring. I'll choose the latter, hoping that my children will be shown ways to love education and explore it like never before.

When you're ready to engage in a lesson on ambiguity, check out *Which One Doesn't Belong* (wodb.ca), curated by Mary Bourassa, for a variety of visual prompts like the ones below.

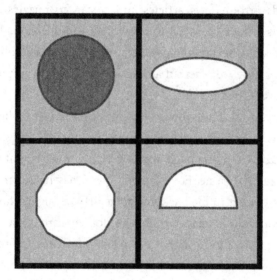

I recently showed this image to my boys. Prior to showing it to them, though, I couldn't help but consider which shape they would choose. Think about it: Which one do *you* think a four-year-old would choose? Why?

I thought my son would choose the image in the top left square—the only circle and the only one which was colored in. Those two things quickly jumped out at me, and I assumed they would obviously jump out to a little kid. Come to find out, I was wrong.

"OK, which one of these four shapes doesn't belong in the group?" I asked him.

"That one," he said, pointing to the bottom left.

After a few seconds of stunned silence, I said, "Now, uhmmm, do me a favor. Tell me ... *why* does that one not belong?"

"It doesn't belong because it's too pointy."

I was genuinely surprised. Did you catch what happened there? My son, the pre-school kid who is barely learning his shapes, just identified the only polygon. He picked it out of the crowd because it was, to him, too "pointy." He saw the only shape with exclusively straight edges, or the only polygon.

When I showed him this same visual four months later, he chose the same shape, but this time gave a different reason: "It has more sides than the other ones."

Once again, he reached for the concept of a polygon without knowing any of its properties or what "polygon" even means. Maybe it was coincidence, and maybe it wasn't. Our brains are wired a certain way; his was fixated on the lower left shape being clearly different than the others.

One outstanding aspect of these prompts is, no matter what your age, there is a discussion to be had. Show the image to your spouse, partner, colleague, or friend and ask them which shape doesn't belong. There is a very good chance you will have a different choice than the person next to you. If not, could you come up with a reason why another one doesn't belong? I bet you can.

With an older child, ask her to pick the shape she thinks does not fit and then challenge her: *Well, you said the top left doesn't belong, but I'm going to argue the bottom right doesn't belong. Can you convince me otherwise?*

The goal of this phrasing is to get the other person to build a defense for her decision. If all we do is seek out correct answers, we never have to build a justification because we never face opposition.

Here's another example:

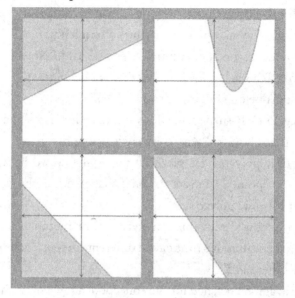

Yes, this looks and "smells" like a problem for a high schooler. After all, each of these images is an inequality, plotted on the coordinate plane. In order to determine which doesn't belong, someone would need to know about greater than, less than, linear inequalities, quadratic inequalities, quadrants . . . Did I lose you yet? This sounds complicated, but it doesn't have to be. Here's what the conversation looked like at our house:

"Hey, boys. Which one of *these* doesn't belong?"

"Top right," Luke pipes up immediately, as if it was obvious. Well, yeah, it's obvious. It's the only curved graph and the only one where the bottom left square is all white.

"Uhmmm, why?" I prompted him, realizing that there may be something that I'm not seeing, but he is.

"Look. It's the only one with color touching only one side of the square," Luke answers, surprising me again.

Did you see it? He didn't see a parabola, any shading, or anything to do with graphs. He didn't get scared off by the inequalities, by the coordinate plane, or by his lack of background knowledge. His eyes went straight to the larger square containing the graph and saw only one side where the shading was touching. Nearly wiping the tears of joy from my eyes, I patted him on the back and told him I liked his response, still processing exactly what had happened.

Throughout my years of *Highlights* puzzle-solving, I was simply searching for the right answers and the feeling of success I got when checking the back of the magazine to confirm my answers, leading to a fist pump of pride. By contrast, my son is looking at a set of images and determining one doesn't belong, based on a certain set of characteristics the others do not hold.

Let me be clear. I am *not* trying to devalue *Highlights* and the exploration of puzzles as a source. I *am* trying to provide another way to look at inquiry and offer up a multitude of options for children to explore and engage with. Additional options will likely introduce children to even more facets of mathematics, invoking more conversation, deeper meaning, and perhaps a little debate along the way.

And, just for fun, the prompts don't always need to be math-related. Check out this post from Mishaal Surti:

What kinds of questions can you ask about these?

Do you have a musician at home? If so, this is right up his alley. Even if you don't, and even if you don't know a lot about music, it's fun to explore what the symbols mean. The bottom left is the only one written in the bass clef, while the top left is the only one with an "accidental" (the sharp symbol by the next to the last note).[1] Bringing a prompt like this to the dinner table could connect your child's interest in music with your attempt to build a bridge in understanding.

The Internet is not the only source of great organizational conversation starters. While setting up the workbench in my garage, I encountered a situation. I was ready to hang tools on the wall, but I needed to remove the spade bits in order to hang their holster.

Both kids were with me in the garage, but the nearly four-year-old was the most interested in helping. Well, why not?

---

1   No, it's not a hashtag. Well, it's kind of a hashtag, but it's been around *much* longer and has a totally different meaning.

"Hey, buddy. I needed to take these spade bits out to hang the holster. Do you think you could help me put them back in order?" I asked. Silence.

Not because he was ignoring me, but because he was looking at the spade bits of various sizes, fanned out in my hand as if I were playing a game of garage Go Fish. He analyzed their different sizes as well as the sizes of the bits I'd already placed in the storage rack as placeholders, as he decided which one he should choose first. After taking inventory, Nolan grabbed the one-inch bit, placed it to the left of the seven-eighths bit, then looked at me confidently.

"This one goes heah, daddy," he said confidently, with his skewed toddler annunciation.

"Hmmm, are you sure? Why do you think so?"

"Betuzz it's bigga than the uvva one."

I let him have the small victory, turning him back to the rest of the bits and watching as he meticulously organized them from greatest to least, filling in the gaps along the way. No, he doesn't know the difference between seven-eighths or fifteen-sixteenths, but I didn't care. I inserted the math commentary along the way so he knew *why* things made sense,[2] but made sure not to stop the flow of conversation just to break down the meaning of a fraction. I wanted him to own the moment and celebrate his growth. This was far more important to me than deconstructing fractions and focusing on rightness.

---

[2] "These two look the same, but fourteen sixteenths is just a little bit smaller than fifteen sixteenths. See? Fifteen is bigger than fourteen, so fifteen sixteenths is a bigger fraction than fourteen sixteenths." Did he soak that in? I don't really know, but I doubt it.

Within five minutes, we ended up with this:

Nolan, a spunky little four-year-old, was grappling with the organization of fractions. If you asked him, he would have said he was putting them in order or maybe "oh-gun-eye-zeeng" the bits.

Take what you're looking at now and transfer it to a number line, something commonplace in an elementary classroom. When children see a number line with arbitrary values placed onto it, there is no frame of reference—no reason to wrap their heads around the concept. However, when a child is faced with a challenge of putting something in order, the task makes sense and they get a feeling of accomplishment and pride when everything looks right.

Truth be told, the one-inch and the fifteen-sixteenths bit look incredibly similar. Nolan switched them up. He initially mixed up the three-eighths and five-sixteenths bits, but the similarities in those two fractions were perfect scenarios for me to hop in and ask questions to help him navigate to the correct solution. If I wanted the answer, I would have told him, "Put the one-inch bit to the left of the fifteen-sixteenths bit because, just trust me, the one-inch bit is bigger. C'mon, lunch is almost ready. You need to go wash your hands before we eat."

Some parents may argue it is easier to engage a younger child with mathematics than an older one. While there is some truth to that because of the ever-changing attitudes of adolescents and the simpler setup of arithmetic, there will always be opportunities to bring math to the table—regardless of the age of the child.

There will always be opportunities to bring math to the table—regardless of the age of the child.

I would be perfectly comfortable putting this same predicament in front of my eighth grade students (thirteen- and fourteen-year-olds). However, I'd likely not include the placeholders and instead give them an empty holster and ask them to fill it up with the bits in correct order. Yes, I'd be there to help answer their questions (and ask some of my own), but I'd also be there to analyze how they set up their process. Nolan did some mental analysis of what was in front of him. An older child might be more inclined to grab the bits automatically and spread them out from least to greatest. But the only way we'll know is to try it out. And there's no better place to do this than at home in a friendly and familiar environment, free of retribution for working through the process.

Generally speaking, fractions are a daunting concept for children to truly understand, particularly when there isn't a visual model like those drill bits to help make sense of the numbers.

Quick! What is 7/8 + 7/32?

See? Even if you got the answer, some cognitive deconstruction had to happen first. If you didn't get 35/32 (or 1 3/32) right away, it's because the fear of fractions is real.

We could spend an entire chapter dissecting the nuance of fractions and *why* we—even as adults—scoff at the very sight of them. However, this book is focused on your having something to talk about with your child, and a deep-rooted mathematical justification of fractions is probably not at the top of your list. However, the work done by Dr. Andy Norton and his team to understand why students struggle with fractions is worth looking into. In his "Zone of Potential Construction" podcast, Dr. Norton discusses the part-to-whole relationship in fractions, among other things. You'll definitely benefit from listening to it.[3]

Bringing it to the Table

As I mentioned earlier, Mary Bourassa has done quite a bit of work to curate and create prompts from all different content areas and all parts of the world. I asked Mary to share a few words about her site and her journey.

From Mary Bourassa, creator of WODB.ca

I believe the great appeal of puzzles is that they are accessible to everyone. My own children enjoy doing them and like to create their own. One of the most powerful features of WODB.ca (the Which One Doesn't Belong website) is that it gives every child a voice. The site lets users look at puzzles in many different ways but does not provide a "correct" answer. The freedom to explore a variety of outcomes allows the student who may struggle in class to find a reason why at least one object does not belong—and it may be a reason that the high flyer in the group did not see. Having all students see multiple solutions fosters a growth mindset and allows for great

---

[3] Brownell, Chris. "Episode 3 | Units Coordination, Its Place and Role in Mathematical Understandings". *ZPC - Zone of Potential Construction*. Podcast audio, September 8, 2016. http://www.aimsedu.org/2016/09/08/ episode-3-units-coordination-its-place-and-role-in-mathematical-understandings/

conversations that promote correct vocabulary and broaden problem solving strategies.

The focus is on *"why?"* Students have to be able to explain their reasoning and make their thinking visible to others. They must all be able to critique their own reasoning and the reasoning of others. They can work quietly or discuss everything they are doing. When they are creating their own sets, they will think that they have a set that works but then discover that it doesn't and now need to fix it. The process of taking apart what they have, deciding what to keep and what they need to add makes them think deeply and demonstrate a solid understanding of the mathematical concepts at play.

There are more simple benefits, too. How do we refer to each object? You could use top left, top right, bottom left, and bottom right, which are important concepts for young children to learn. Older students can refer to the objects in a more mathematical way by using quadrant numbers. These sets can also be used to motivate new vocabulary such as quadrilateral, rhombus, parallelogram, isosceles triangle. Children find their own way of describing what they see and often don't realize that they are saying the same thing as another child using different words. The language development that can be tapped into is a powerful tool.

The website has WODB sets separated into shapes, numbers, and graphs, but also includes a few incomplete sets. These are sets that have two or three items that need to be completed to make a full WODB. I love how many different solutions there are to these and how much discussion arises from them. One person will suggest adding an item to complete the set, and another will discover that it means that another item no longer works so they have to try again. Perseverance will pay off here!

Although the WODB sets on the website are mostly math-focused, they can be created around any topic or use any objects you can find in your house or even outside. Children will be developing their critical thinking skills and having conversations where they must justify their ideas and listen to others' ideas. And they will have fun while doing so!

Encourage

- Everyone to listen to others' ideas
- Everyone to find a reason for each item not belonging
- Making connections between reasons
- Everyone to dig deeper and find multiple reasons

Avoid

- Stopping at one right answer as there should be many correct choices
- Valuing one reason more than another
- Seeing a prompt as too difficult or challenging for your child

7

Fraction Talks

S peaking of fractions, let's talk about that dirty "F-word."

Tamara, Jason, Aaron, and Diana are sharing a pizza. If the pizza has twelve slices and they split the pizza evenly, how many slices will each person get?

Oh wait!

This problem was in *your* junior high math book, too?

If not this exact problem, do you remember something eerily similar? These problems are a classic way to bring the "real world" into a math problem. The "real problem," though, is that questions like these also push kids away. No matter how you slice it, there are only so many times we can use pizza to represent fractions before students get sick of it. Instead, we need to talk about fractions in creative ways to spark curiosity—not only in ways involving food. Here are a few examples:

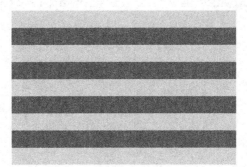

What fraction of the flag is made up of each color?[1]

---

[1] The light shade is yellow and the darker shade is red.

"OK Luke, how much of the shape is colored red?"

"Four."

"I don't understand. Can you explain?"

"There are four red stripes."

"Ah, I see. So how many yellow stripes?"

"Four. Oh wait, FIVE!"

"So how many total stripes are there?

(waits, counts in head)

"Eight... no, NINE!"

"All right, so how much of the shape is red?"

"Four... four out of the nine."

"Very good, son."

While it may seem trivial to some, this is a task involving fractions. The flag of Catalonia, shown above, contains nine evenly spaced horizontal strips of color. Four of these strips are red and five of the strips are yellow. Without a doubt, we can say the flag is four-ninths red and five-ninths yellow. At six years old, my son was able to have an in-depth conversation about fractions that didn't scare him away. Think about the value of this conversation as more of them happen between now and the time he is formally met with fractions in his mathematical development, creating less of a stigma and more of a puzzle based on images that are accessible and approachable.

Catalonia provided us with a nice warm-up. Let's try a bit more challenging one.

What fraction of this flag is made up of each color?[2]

"How about this one; how much of the flag is blue?"

"One. One out of three."

"Hmmm. It looks like there's something different, though. What's different about these colors?"

"Well, the yellow one is bigger than the blue and the red stripes."

"Ah, that's what I was missing. So I'll agree with you that one out of three colors is blue, but it seems like there's more yellow than the others. Let's look at it like this:

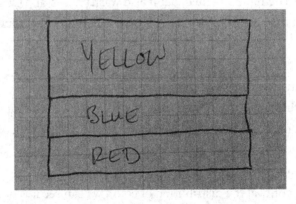

"Does it look like one out of three boxes are going to be blue?"

"No."

The brevity of his reply was due more to intrigue than confusion; *wait, something doesn't look right.*

---

[2]   The light shade is yellow, then blue and red.

"Let's make this a little bit easier. Look what happens if I do this:

NOW how much of the shape is blue?"

And because we had just talked about the Catalonian flag, my son brought in the same language to the Colombian flag, declaring that "one out of four of the shape is blue." It certainly helped that I had graph paper handy, but it wasn't necessary. It also helped that the shape lent itself to being subdivided, but it also wasn't necessary. Once again, we were able to have fractions serve our conversation without writing down a single number.

To reflect on this without the dialogue:

Some may see their solution immediately. Three different colors make up this flag of Columbia. The yellow appears to take up half the flag, while the blue and red appear to make up the other half. However, this creates a bit of a problem. Since there are three colors, shouldn't the denominator be three?

Looking only at the distribution of color, we can see it is possible to split the flag into four equal parts. When we do this, we see the yellow occupies two of the four parts, while the red and blue each occupy one of the four parts. Therefore, the yellow is two-fourths (or one-half) of the flag, the blue is one-fourth of the flag, and the red is also one-fourth of the flag.

Fortunately for us, there are hundreds of flags we can use to challenge ourselves and everyone else around the dinner table. Take a look at some of the others requiring a bit more paper in order to discover the solution:

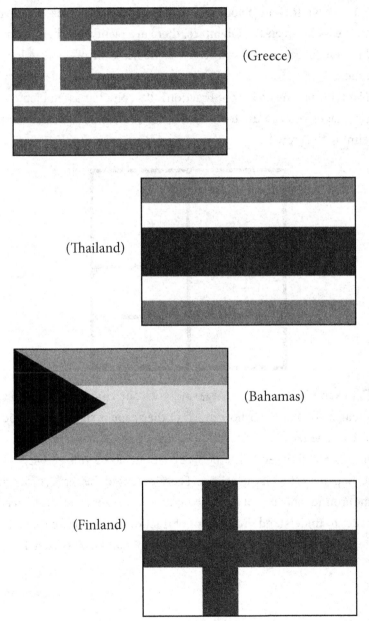

(Greece)

(Thailand)

(Bahamas)

(Finland)

If it's an Olympic year, this fraction activity is an easy sell. Watching the Parade of Nations as athletes march into the arena waving their flags proudly gives this "flag challenge" relevance and gives the people at your table something else to think about besides how much longer they have to wait before the Artistic Program begins. However, in the years between the Olympics, there are plenty of other prompts at FractionTalks.com for you to use to spark a lot of good, relevant questions.

Nat Banting, the creator of FractionTalks.com, has assembled quite a few images you can use to bring fractions to the table without overwhelming the crowd.

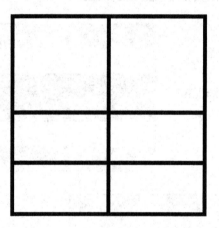

For example, draw this image and ask your child to color in a set of squares. Then ask her how much of the shape is shaded. You know your child better than I do, so I'm assuming you will phrase the question to her ability level. The best part about doing this activity on a sheet of paper is that it can change. Feel free to fold the shape, move the shape around, or even cut the shape into pieces. What your child needs in order to understand the concept of fractions is the desire to manipulate them; you can provide the piece of paper and a bit of sketching.

Learning fractions is *not* the goal of these challenges. The goal is to engage your child in the process—to unravel the puzzle and understand the representation. Yes, we want to know the fraction of the shaded area or the fraction of the flag's colors, but we *really* want kids to be curious about something they've seen before but never thought to ask questions about before this moment. If your child talks about shapes and wonders about their color distributions, you'll know your fraction talks have made a big impact on him.

Here are a couple more examples from the site, although you really need to head over and take a look for yourself:

What fraction of the shape is shaded?

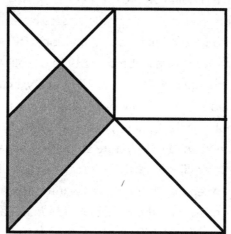

What fraction of the shape is white?

Bringing It to the Table

Look anywhere and you can see an image that has been broken up into fractional pieces. Take a look at what Nat Banting says below about structuring your fraction conversations and keeping the conversations going.

From Nat Banting,
creator of fractiontalks.com

I created fractiontalks.com to fill a need in my classroom. It seemed like every year my ninth-grade classes were filled with students that had stopped thinking about fractions. This is not to place blame on the students or ascribe some type of collective lethargy to a younger generation, because, in all honesty, I couldn't blame them. School had systematically done away with thinking about fractions and replaced the thinking with execution. Many times this transition was performed long before students were able. You may experience the same dilemma; there just doesn't seem to be much to talk about with regards to fractions. I wanted something simple, familiar, and peculiar: something that would beg for an answer but not readily make one available.

The images you see throughout this chapter (and on the website) are common to classroom textbooks,[3] but textbook versions rarely, if ever, employ the use of peculiar sectioning to sponsor thinking. In textbooks, sections almost always appear the same size—pre-sectioned. A fraction talk image can contain sections of many different sizes resulting in an image that is familiar enough to structure conversation but unique enough to facilitate wonder. They often result in familiar shapes like squares, rectangles, and triangles, but can

---

[4]   In math education circles, they are known as part-whole area models.

also include less familiar shapes like trapezoids, rhombi, and even circles (if you're brave).

This book is aimed at providing starting points for mathematical conversations, and there is one overwhelming starting prompt begging to be addressed: *How much of the square is shaded?* There are a couple important things for you to notice about this prompt. First, it doesn't include the word "fraction". This is intentionally done to distance formality and allow it to form as the conversation dictates. Second, the question is versatile and can be re-used for many different types of shadings and for many different images. It grants each image staying power.

However, every prompt gets stale, and it is critical that conversations remain interesting. There are two fundamental ways to open fraction talks to allow more space for conversation: change the prompt or change the image.

Changing the prompt is often instigated by the conversation itself. Prompts are in no way fixed, and should be posed to fit the conversation as it unfolds. Fractiontalks.com provides some examples of prompts that could be used under the *How to...* tab. There really isn't a bad prompt to start conversation as long as it provides space for children/students to operate as they see necessary. Consider the progression of prompts below:

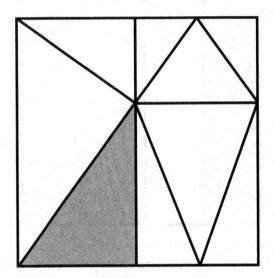

- How many ways can you shade more than one-half?
- Find all the sections that are the same size as the shaded one?
- Can you shade two sections that are still smaller together than the original shaded section?
- Can you shade three sections that equal exactly three-quarters of the whole?

Each of these provides space for student choice and justification—two critical elements of conversation. Each also offers a varying level of sophistication with the topic.

Changing the shape does not always simply mean presenting another fraction talk image (although this may be a profitable course of action in the moment). Consider how the thinking might shift if the image was altered so as to no longer create "nice" sections. For instance, use the shading in the left-hand square to justify how much of the right-hand square is shaded.

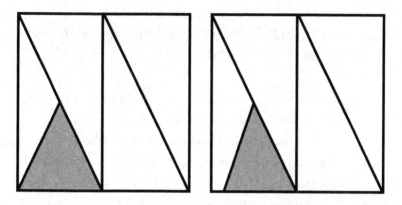

Further shifts in conversation may be sponsored by changing the "whole" from a square into another shape.

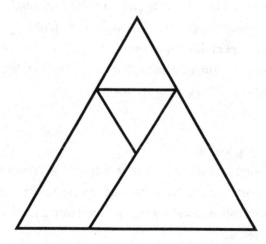

The goal here is not to teach formal fraction operations, but, rather, to find curiosities that lay the groundwork. It is to allow the conversation to form subsequent wonders and not to provide a prepared sequence of action.

Encourage

- Justifications
  - » Some of the simplest mechanisms for moving a conversation forward are asking, "*What did you see?*" or "*How do you know?*"
- Creation
  - » Great conversations often emerge when they are inspired by a personal creation of the child/student.
  - » Mark on the diagrams, add new lines to make new sections, and pose new prompts. This allows conversations to move forward on the back of "*What if?*" statements (like, "*What if I added this line?*" or "*What if I wanted to shade double?*").
- Comparison
  - » Words like "*bigger/biggest*," "*smaller/smallest*," and "*equal/identical*" begin to establish the notion of magnitude.
  - » Prompt children/students to find all sections of the same size or estimate a value that is "*too high*" or "*too low*" if they have trouble getting started.

Avoid

- Rushing to notation
  - » Notation isn't bad if it is used to support conversation. Insisting on notation throughout the process may transform an exploratory conversation into a mechanical exercise.
  - » As it becomes necessary, tether the conversation to fraction notation as the child/student is comfortable.
- Answers as goals
  - » Conversations are about ever-emerging possibility; therefore, focusing on an end result might stunt great thinking. Keep the conversation rolling with questions like, "*Can you see it another way?*" or "*What other question could we ask?*"

8

Talking About Math

"Starting at 4 and counting by 7, how high do you think we can count?" Mom asked.

My parents had a weird way of keeping us entertained on our long drives to Michigan every summer. Yes, we counted signs and played "I Spy" throughout the four-day excursion, but you can only play those so many times before your little brother blurts out imaginary things just to annoy you and make you cross your arms in protest. So our parents came up with creative ways to switch things up and make our brains hurt a little. For example, take this conversation we had when I was nine and my brother was seven.

"Eleven. Ha! I win," I shouted.

To which my brother added, "Not fair!" And then after a little thinking time, "Eighteen!"

"Twenty-five," Mom called out the next number in the sequence.

"Thirty," said Dad.

In unison, my brother and I cried out, "*Noooo! That's not right!*"

My father, an engineer and pretty good with numbers in general, was always quick to play the struggling student. We always suspected shenanigans when it was his turn, and I now realize how positive this was for me. Rather than taking my parents' answers for granted and giving them more value than my brother's or my own, I analyzed everything he said and checked his work in the process.

We were participating in what Sadie Estrella, a teacher, math coach, and lesson writer in Hawaii, has called a "Counting Circle." In classrooms, a room full of students might stand in a circle and, from a starting value, count up by a set increment. The teacher designates a couple of students to give incorrect responses and, while everyone knows this, no one (besides the two selected) knows who will intentionally be calling out the wrong numbers. Everyone listens carefully and quickly checks each other's work to ensure the circle progresses as it should.

The first benefit of a Counting Circle is that it actively engages all the students. Similar to what my brother and I did in our car, the check-and-balance occurs naturally. We wanted to make sure everyone was right, so we paid close attention. In a classroom, the last thing students want is to get derailed off their intended path—especially when they know for a fact someone has been told to give the wrong answer. And this leads us to the next benefit of this activity.

The second, and possibly bigger, benefit of a Counting Circle is the expectation wrong answers will occur. Often, getting the wrong answer carries such a strong negative connotation. In what we assume is a "normal" traditional educational experience, we are praised for right answers; wrong answers are chastised and ridiculed, even punished. In this activity, however, students are encouraged to correct errors, work together to determine where the mistake was made and why, then move on once the error is rectified—just like my dad modeled for *me*.

"Okay, fine. I see my mistake. Thirty-two," Dad corrected himself.

"Thirty-nine," I said.

"Forty-six," my brother added.

And so on—until Mom taps the brakes on the activity: "Now let's start subtracting by four. Ready? One hundred five."

As I remember it, we took *fifty hours* counting down to zero. In reality, it was probably closer to five minutes. But the game provided a useful discussion in a seemingly unending car ride, plus it gave us all a chance to flex our mathematical dendrites[1] when we wouldn't have been doing much of anything else otherwise.

The game provided a useful discussion in a seemingly unending car ride, plus it gave us all a chance to flex our mathematical dendrites.

This same activity works well around the dinner table as well. Think about it: How many times have you sat down to dinner, only to discover no one has anything to talk about? Stirring up a quick counting circle is an easy fix for this scenario. Plus, it can accommodate any level of learner. For example,

One, two, three . . . (count up by ones).

Zero, two, four . . . (count up by twos).

Negative three, negative nine, negative fifteen . . . (count down by sixes).

Four, nine, sixteen . . . (count up by perfect squares).

Once again, it's perfectly fine not to know the answers. Being vulnerable and open to learning alongside your child will show her mathematics is something to be worked on over time—not just something to write down on an exam. While it is not true we all use math every day, it *is* true we can all apply mathematical concepts and conversations to our daily routines.

In the classroom, a convenient way to demonstrate vulnerability is through a "Number Talk." In a nutshell, number talks are brief conversations about the solution to a mental math problem. If you're a teacher, I highly recommend grabbing a copy of Cathy Humphreys' and Ruth Parker's work, *Making Number Talks Matter*. I have incorporated

---

[1] Dendrites are those synapses in our brain that tingle with joy every time we learn something new or think hard about something. The more dendrites firing, the better!

Number Talks in very simple ways during workshops I've done or classes I've taught. An example might be:

Using only the tool between your ears, think about five times nine.

What is the answer?

Now tell me how you got your answer.

Good!

Now, to make things a bit more challenging, what is twelve times seven?

How did you get your answer?

Some people, no matter their age, will say they "just knew it in their head" and they would more than likely be right. Dig a little further, though, and you discover something great going on.

What do you mean that you just knew it? What did you mentally do in order to answer that question?

The following are some responses I received from a group of elementary students:

I multiplied seven times two, then seven times ten, then added both together.

I multiplied seven times ten, then added seven to seventy, then added seven more and got eighty-four.

I multiplied seven times six and got forty-two, then multiplied forty-two by two and got eighty-four.

So, did you use any of these methods—or did you find a different one?

Each student who responded above had a unique system of patterning. Each took the number twelve and deconstructed it to meet their comfort level. This example can work for any type of problem your child is currently comfortable with and can calculate in her head (though a calculator is handy for checking the work). You might ask . . .

What is thirty-two plus nineteen? How did you know?

What is fourteen times eight? How did you know?

What is one hundred thirty-three minus forty-five? How did you know?

What is the square root of eighteen times the square root of twenty?[2] How did you know?

Again, the focus is on the reasoning behind the solution and not necessarily the answer itself. Hiding the answer from someone trying to engage in a Number Talk can be counterproductive to understanding her thought process and, thereby, create unfortunate habits. Because we are putting so much emphasis on the process of the problem, allowing children to confirm the solution gives them a chance to validate the work they did to reach a conclusion. Without this closure, there is no room to investigate errors and reflect on the steps it took to get there. Maybe the key is actually sharing the answer early on.

What is thirty-two plus nineteen?

Fifty-one.

Yes, you know it is fifty-one because your phone's calculator says so. But *why*?

Letting your child use a calculator can make numeracy more appealing and approachable. And getting the answer from a cell phone calculator isn't cheating; adults do it all the time to get through their normal routines. What's important is being able to interpret a solution and explain it.

Whether it's a counting circle, a Number Talk, or huddling around a calculator, there are critical ideas that can be shared when we sit down with our child to talk about numbers.

Bringing It to the Table

Out of all ideas discussed throughout this book, having a conversation about nothing more than numbers is the most accessible and easiest to adapt to your child's ability level. When I'm talking with

---

[2] A possible explanation is this: Since they are square roots being multiplied, you can multiply eighteen by twenty to get 3,600. Take the square root of that and you'll end up with 60. If you're still stuck, email me and I'll help.

my four-year-old son, we start at three and count up by ones. When I'm talking with my six-year-old son, we start at two and count up by threes.[3]

As a parent who is also a lover of mathematics, my fault in bringing these conversations to the table has been getting too challenging too fast. While the prompt is accessible for each of my kids, I need to be cautious about weaving in too many hurdles along the way.

While working with my middle school and high school students, the challenge was not making it too challenging. A junior might be enrolled in an honors math class, but there's something novel about jumping into a circle where they start at nineteen and go up by seven. Or begin with 35 and subtract 8. Or . . . you get the point.

This activity by itself would be a good way to kill some time and get the wheels turning. Pair that concept with the breakdown of *how* your child got to a certain number, and the light bulb *really* lights up.

Starting at four, count up by four.

"Eight."

"Twelve."

"How did you get to twelve?"

"I put the eight in my head and counted four more."

Sure, a simple interaction but a very important one for my six-year-old to explain as he develops an understanding of numbers.

Encourage

- Taking a wild guess, when it seems appropriate
- Finger counting
- Taking whatever time they need

Avoid

- Using a calculator
- Blurting out answers on your child's behalf
- Getting to a point where mental calculations are no longer possible

---

[3] Just as one example. We use others, but this one gets the most discussion.

9

Noticing and Wondering

Take a moment to look at the picture on the previous page. Then, respond to both of the following questions:

What do you notice?

What do you wonder?

To be fair, a lot of *noticings* and *wonderings* have very little to do with math. My brother took this picture of me as we headed through the upper peninsula of Michigan on a warm August afternoon. When I shared it at a workshop, it had an entire room full of adults not only noticing and wondering—but also questioning and engaging in debate. (And just for the record, I'm actually 5'9"—not 4'10" as this picture suggests!)

Some of their *noticings* may sound a lot like your own:

That's a lot of snow!

I see the low was only 116 inches. Wow!

Over a fifty-year period, the average is about twenty feet of new snow.

Looks like John would have a hard time getting out of a Yooper[1] snowstorm.

Some of their wonderings included the following questions:

How quickly did the snow melt between new snows?

How much water did all the snow make?

Were people able to commute during the record snow season?

How many trucks and people did it take to clear the roads during the snow season?

Why would anyone want to live up here?

---

[1] A Yooper is "a native or resident of the Upper Peninsula of Michigan," according to Merriam Webster. There are plenty of jokes and tales about Yoopers worth exploring. Google them and enjoy.

Any of those wonderings—none of them created by me—would have made a fantastic launching point for a discussion. But for our group, the winner was how much water the snow made. Knowing the Internet was a far better source than our imagination on this one, we went there. According to theweatherprediction.com, the "snow to liquid ratio is 10:1. This is saying that if 10 inches of snow fell and that snow was melted it would produce 1 inch of liquid precipitation in the rain gauge."[2]

Once we had gathered this little nugget of awesomeness—later verified by the National Oceanic and Atmospheric Administration (NOAA)—we were off to our calculators to determine exactly how much water was generated by the 390.4 inches of snowfall. But we didn't stop there because now we were curious about what happens when thirty-nine inches (or over three feet!) of water are dropped into Lake Superior.

As we continued our discussion around the table, the questions from this group of adults became larger than we could—or cared to—answer. Looking back, the curiosity sparked by this single image is something I'll always remember.

"What do you notice?" and "What do you wonder?" are concepts formalized by Annie Fetter and The Math Forum, now a part of NCTM.[3] However, people have been asking these questions informally for as long as they have been put into scenarios where decisions need to be made. Whether math-based or not, I'm sure an image has caught your attention and made you wonder. Perhaps a billboard didn't seem quite right or you noticed something awry with a stack of books. In those moments of perplexity, we can unravel the observations and questions necessary to reach some sort of resolution.

---

[2] Jeff Haby, "Snow To Liquid Equivalent," *Haby's Weather Forecasting Hints*, accessed March 24, 2017, http://theweatherprediction.com/habyhints/346/.

[3] The National Council for Teachers of Mathematics, or NCTM, is dedicated to providing positive math interactions for students around the country (and beyond). Check out some of their resources at nctm.org.

Consider ... moving day.

Disclaimer: I despise moving. When I was a kid, we moved more than we wanted, but thankfully, I was a kid for the majority of our moves, so I got a free pass from lifting the big stuff. Unfortunately, now that I'm an adult, the heavy lifting is on me. So when it was moving time, I found myself in one of those "moments of perplexity."

Based on the image above, I'll ask you the same questions again and the ones I asked my wife:

What do you notice?

What do you wonder?

This is a nineteen-foot travel trailer designed to haul dirt bikes, four wheelers, and other camping supplies. Instead, we were loading it to the gills with heavy boxes full of memories and memorabilia and furniture and furnishings. My father-in-law generously lent it to us with the caveat it should handle somewhere in the ballpark of 1500 pounds. No problem. Sure.

Then I started making some observations:

I have a ton of space.

I am leaving plenty of room in the middle.

The trailer is starting to get heavy—fast!

I still have to do something with all the extra space before driving it twenty miles on the freeway.

Yikes.

And then I started asking questions:

How many more boxes can I fit in the trailer?

Wait. How many more boxes can I safely fit in the trailer so the tires don't explode?

What is the total weight right now?

Where can I find light enough supplies to fit into the gap?

Why does my back hurt?

All these questions (especially the last one) ran through my head as I called my wife over, grimacing because something wasn't right.

My wife and I stared into the travel trailer with mental weight scales, estimating how much each bin weighed and whether or not we could reach our destination without ruining her father's trailer. We estimated we were right around 900 pounds, dangerously close to the limit considering the left tire was showing some signs of stress.

Not only were we playing the ultimate game of moving-day Tetris,[4] we were also trying to balance two sides of a trailer without going over the weight limit *and* doing our best to keep boxes from tumbling during the commute. Had our kids been a bit older, I would have followed

---

[4] Wait, you don't know what Tetris is? Head over here: freetetris.org. You're welcome—or I'm sorry—in advance.

my parents' lead from family hockey night so many years before and turned this "unfortunate situation" into an opportunity for the family to work together to solve this problem. They would have been invited into the "game" to help figure out the best configuration, instead of playing in the backyard so they didn't get crushed by one (or more) of the boxes. But finding the right time and place for math conversations is critical, and this was neither for a three and five-year-old. Later, they could marvel at the puzzle we managed to finagle before making the trek to our final destination.

Imagine approaching this giant tower of pumpkins. Now, answer the same questions again:

> *What do you notice?*
>
> *What do you wonder?*

My immediate question was "How long did it take them to make that?" Yours might be:

How many pumpkins are in the tower?

How many pies could you make from all those pumpkins?

What would happen if I took the one at the top?

What is the value—in pumpkin dollars—of the tower?

Most, if not all, of the questions raised will have something to do with math. Explore them, play with them, and be okay with the conversation getting a little messy.

Bringing It to the Table

On top of the fact that Annie Fetter has built a career around the ideas of noticing and wondering, I have a profound amount of respect for the work that she does in developing teachers' and students' vocabulary with mathematical curiosity.

Annie Fetter, co-founder of The Math Forum

I love that John has included *notice* and *wonder* in this book. When I saw the photo of John and the snowfall depth stick, I noticed and wondered a lot of things really quickly! Noticing and wondering are things that humans do by nature. And it's beneficial and rewarding to show your child that you value his or her observations and curiosity. Since we started doing this with students and encouraging teachers to do it as an activity, those teachers and we have learned tons about what students know and what they're curious about.

In a five-minute talk I gave about noticing and wondering (which you can watch on YouTube[5]), I share a story of a three-year-old boy and his father who were on an airplane with me. The father noticed

---

[5]   tabletalkmath.com/Fetter

the way I asked his son questions and really dug into his thinking. My approach and the child's response were apparently both very different than what happened when the father just answered his son's every question. He vowed to talk to his son differently from then on.

As parents, we often reflexively answer questions our kids ask instead of first asking them what they're thinking, noticing, and wondering. We all like our ideas to be heard and valued, and children are certainly no different. The next time your kids ask a question, ask them what they think instead. The next time you see something cool that piques your curiosity, ask your kids what they notice about it. Telling your children about something? Ask them if there is anything they are wondering about. Wondering, in particular, is a very purposeful choice of words because wondering is a different thing than asking questions.

Children have ideas about almost everything, but often think they can't tackle something or dig into it because they don't know "the way" to do whatever it is. Asking them what they notice and wonder and helping them develop curiosity as a tool they use on their own will help them get started on just about everything. Much of the time, getting started is the hardest part.

Last fall, I got some help at a home improvement store from a man, Junior, who had dropped out of school several times but eventually opened his own barber shop. We talked about what I do, and he wondered (honest!) what he might do to encourage his son to stay in school and make sure he had more options available to him than he'd had. I suggested ways he could encourage his son to think more mathematically more frequently and ways in which Junior could elicit and value his son's mathematical ideas. One of those was to frequently ask his son what he noticed and wondered and to engage in that as a family activity. Adults notice and wonder all the time, and it's great to share that often hidden thinking with your children. While I don't have Junior's contact info to follow up, he seemed eager to try it and see what happened. I hope he did!

Encourage

- Your children to share their ideas
- Asking questions you don't know the answers to
- Seeking the answers to your child's questions together

Avoid

- Telling them pretty much anything, especially how they should think about a particular situation
- Asking questions you already know the answer to
- Praising them for being fast or even right, unless they explain their thinking

10

Leveraging Technology... or Not

Most mornings, my wife and I don't hear a word from the boys. Before we were parents, this was perfection. Nothing to hear meant there were no problems, just bliss. With little ones in the home, though, the sounds of silence can mean trouble. When our youngest was two, he would come up beside our bed and tap my wife on the shoulder around 5:30 a.m. and whisper:

"Mommy?

"Mommy. Wate up.

"Mooooommmmy, wate up!"

Confident she just needed a little more encouragement, he would use his index finger and thumb, spread her right eyelid open to expose what was a perfectly peaceful eyeball one second earlier, and gleefully announce:

"Mommy! You're awate!"

Writing about this makes me giggle—now—but, at the time, it was frustrating to live through. My response (usually silent) was a plea for more sleep, "Kid, you're awake before the sun is! Go back to bed!"

Lately, in the mornings, we don't hear a peep from the boys. Nothing. We wake up on the weekends, and the only sound we hear comes from our slightly off-balanced ceiling fan that makes a faint, repetitive squeak. Nothing more. Strolling down the stairs, we see two boys glued to the family iPad or my wife's iPhone, watching yet another episode of *Fireman Sam* or *Wild Kratts*.

The use of technology in our home is more of a novelty than a learning tool because I'm not convinced an app, no matter how entertaining it is, will help my kids become better academics. Listening to the boys talk about what they learned from watching an episode of *Doc McStuffins*, *Dino Dan*, or *Wild Kratts* is much more appealing to me than seeing them locked into yet another game proclaiming to raise kids' test scores in mathematics.

I'm not trying to de-value the use of technology to support the growth of math fluency at home; on the contrary, I think it can certainly be used effectively. However, I am trying to avoid relying on it as a tool to put my children into a competitive advantage over their peers. If there was an app to do this, we would have heard of it by now and everyone would be using it.[1]

Game On!

The amazing people at MathMunch.org have scoured the Internet and created a list of online mathematical games. While it's still a good idea to personally preview the game to ensure it is appropriate for your child, I wholeheartedly trust the work of Anna, Paul, and Justin. Head to mathmunch.or/games, and enjoy! While you're at it, peruse the rest of their website for art projects, math tools, videos, and more. You won't be disappointed. If you are an educator, this is also a gold mine of ideas to switch things up in your classes!

---

[1] As of this writing, "that app" does not exist. For the sake of humanity, I hope it never does.

The best "app" our children can access is quality interaction with someone at home. Throughout *Table Talk Math*, the goal has been to offer suggestions and prompts to spark an idea, a thought, or a discussion. Yes, there are apps, websites, and programs providing interactivity and discussion, but it is built on artificial intelligence. No matter your educational background, sharing *your* intelligence with your child creates the strongest bond and best opportunity for retention.

The best "app" our children can access is quality interaction with someone at home.

We cannot, however, ignore the large role technology plays in our child's life, no matter their age. As such, here are some tips to get the most out of a phone, a tablet, or a computer, and to use technology as an ally rather than a nemesis:

- **Check in.** Rather than assuming that your child is doing something productive on his device, check in with him every now and then to find out what he is learning. With my boys, I press the pause button on the iPad and ask them to teach me what they have learned so far.

- **Make it a team effort.** Games like 2048 and Cut the Rope have sincere roots in mathematics and can often be explored by a (small) team. Yes, it's probably true those games have lost their luster to newer, better apps, but the premise remains. In games where there is strategy, work with others to come to a solution. Taking turns is great, but if there is no interaction between hand-offs, no real connection is being made and the moment of reflection is lost for everyone.

- **Become the pilot.** Related to team effort, take control of the game from time to time and let your child be the navigator. Ask her, "Where should I move next? What number do you think should go here?" By asking the questions you would internally use to proceed, you are drawing out the thought process of your child and inherently initiating a conversation which otherwise would have remained in her head. Because your child will want to play, too, this won't be a strategy to use one hundred percent of the time, but it is one of many ways to encourage a collaborative spirit.

- **Put a lid on it.** Putting a limit on your child's device time is not just for the child; it's also for the guardian. As new parents, we let our kids get onto the iPad, and time would slip by. Two hours later, the boys were still on the device. We don't blame the kids; the programs they were watching and the games they were playing were designed to keep kids engaged. Once we realized *we* needed a timer, we would set one for thirty minutes. Now, when we hear the timer, everyone knows to shut it down and recap what we learned. The time is not hard-set; some days there will be more screen time than others. What *is* hard-set is the reflection afterwards and the chance to come together and discuss what was learned.

- **Do your own quality control.** Rather than trusting anything asserting to engage your child, run through the program on your own to ensure the validity of its content and the message it claims to convey. With the number of companies creating software to support your child's (and your) mathematical fluency, the market has become flooded with products, but few will be what you are looking for. It may only take three minutes to run through and vet, and it will be time well-spent.[2]

---

[2] Look, no parent wants to be the fun police, but I'd rather include this and be safe than not include it and regret it. Seriously folks, it's worth the time.

By the time you're reading this, your child might already have an interactive watch,[3] a cell phone doing math for him,[4] and a tablet commanding his attention every waking moment. Even as adults, we are guilty of adhering to the noises and interactions on our screens more than we should. As such, it is even more important to set them aside for what matters most: the connections we have with the people in our lives.

---

[3] Yep, my four-year-old got one for his birthday... leapfrog.com/en-us/products/leapband.

[4] It's real. And it's free: itunes.apple.com/us/app/photomath-camera-calculator/id919087726?mt=8.

11

Conclusion

Parenting is hard.

You already know this, but it's absolutely worth mentioning. My wife and I see how tough it is to navigate the role of parent as we weave in and out of life's obstacles with two young kids. We are acutely aware as the kids get older, those obstacles only grow bigger, more stubborn, and more necessary to address. We also do this with the full understanding that we are in a dual-income family, whereas many others are doing this as single parents, a parent who shares their child with a separated partner, or a variety of other circumstances.

This is not the first book to encourage parents to help their kids develop and enhance a love of mathematics. However, my hope is you found something in this book—an idea, a website, a professional mentor—which sparked a thought and provided an insight you did not have before reading *Table Talk Math*. Selfishly, writing this book helped me develop a framework for working with my own children to bring math to the dinner table—without it feeling like math at the dinner table.

I want my children to grow up and be curious.

I want my children to grow up and be confident.

I want my children to grow up and love a challenge.

I want my children to grow up and know it is safe to take a risk.

Maybe those are the aspirations of a parent with young kids. But I'm sticking with them.

Thank you for taking time to join me at the table to discuss math. You'll find an up-to-date list of ideas and resources at tabletalkmath.com/resources, and you can sign up for my weekly newsletter at tabletalkmath.com. The newsletter shares and explores websites, ideas, and tools and includes guest posts from other math educators. For a list of previous newsletters, go to tabletalkmath.com/previous-newsletters. Feel free to click around while you're there.

As you get ideas or stories to share about your own *Table Talk Math* conversations, I'd love to hear them! Email me at john@tabletalkmath.com or find me on Twitter via @TableTalkMath or @jstevens009.

Additional Resources

TableTalkMath.com

Head over to the website for a list of free resources you can use right away with your child. All websites shared in this book, and many more, will be hosted there and updated regularly. In addition, here are a few other incredible sites worth exploring:

Classroom Chef
classroomchef.com/links

Matt Vaudrey and I built this website (and wrote the book, *The Classroom Chef*) to support teachers by incorporating new and creative ways to engage students with math and to make them curious. The resources shared in this book, along with *a lot* more, are stacked up on this link and ready for you to use immediately.

Math At Home, CMC
cmc-math.org/mathathome-2/

This great PDF contains a long list of books suitable for kindergarten through fifth grade students and includes references to mathematics.

YouCubed
youcubed.org/parents/

As mentioned earlier, Dr. Jo Boaler and her team at Stanford University have created this site to support students of all ages learning about math.

How Parents Can Help Their Kids Learn Their Multiplication Facts
mathfireworks.com/2016/09/parents-can-help-kids-learn-multiplication-facts/

From skip-counting to questioning strategies, Tyler provides in his blog post a comprehensive list of ways you can raise your child's numeracy competence in an easy-to-digest post.

PBS Parents
pbs.org/parents/education/math/math-tips-for-parents/math-books/

If you're looking for a list of updated books for kids with math woven into the pages, take a look through these and let me know what you think of them. (As of this writing, I'm purchasing *Clocks and More Clocks* and *How Much is a Million*?)

Bedtime Math
bedtimemath.org

Whether you're wanting a parental blog post, a fun task, or an archive of previous posts, *Bedtime Math* is a fantastic place to visit when you are looking for something new. The team of eleven has curated content from across the Internet, created its own, and made it accessible to parents everywhere.

Daily Desmos
dailydesmos.com

If you and your child do not yet know about Desmos and its power, go to desmos.com/calculator and mess around. If you're looking for some mind-blowing representations of math, check out desmos.com/math or beautiful art projects made of nothing but functions at desmos.com/art.

For a ramped-up version of table talk, go to dailydesmos.com. Built by Dan Anderson and open for guest contributions from all over the world, *Daily Desmos* is a healthy dose of obscure graphs and head-scratching challenges for your upper level child. Beware, though: these graphs can get pretty tough to decipher. Fortunately, many of them have been "solved" by others and the solutions are in the comments section, when available.

Math Books
tabletalkmath.com/book-list

Tons of books support numeracy with your child, and I certainly appreciate your taking the time to invest in this one. To make further reading easier, I have put together a list of books I feel would be good to add to your bookshelf if you are interested in going deeper into ways you can engage your child in mathematical discourse. The website will be updated regularly as new books get published or shared with me so I can pass along the goods to you.

Contributors

The following educators helped make *Table Talk Math* what it is. Simply put, this book has very little meaning and impact without the work of these people from whom I have learned so much. Something I want to make abundantly clear is the respect I have for the #MTBoS community and these six influencers in particular. If you are interested in continuing the conversation with them, bringing them to your child's school, or contributing to their projects, please reach out to them.

Christopher Danielson (@trianglemancsd) - Foreword

Christopher has worked with math learners of all ages—twelve-year-olds in his former middle school classroom, Calculus students, teachers, and children of all ages at Math On-A-Stick at the Minnesota State Fair. Find more of his writing at his website: talkingmathwithkids.com.

Andrew Stadel (@mr_stadel) - Estimations

Andrew is the founder of estimation180.com and currently serves as a digital learning coach for teachers in Southern California.

Fawn Nguyen (@fawnpnguyen) - Patterns

Fawn is the founder of visualpatterns.org and currently teaches middle school math in Southern California.

Mary Bourassa (@marybourassa) - Organization

Mary is the founder of wodb.ca and currently teaches high school math in Ottawa, Canada.

Nat Banting (@natbanting) - Fraction Talks

Nat is the curator of fractiontalks.com and a mathematics teacher in Saskatoon, Saskatchewan, Canada.

Annie Fetter (@MFAnnie) - Noticing and Wondering

Annie is one of the founders of The Math Forum and currently works with the National Council of Teachers of Mathematics.

Image Credits

Chapter 4—"Large Roll," page 28; "Toilet Paper Packaging," page 30; "Smaller Roll," page 31; "Weights," page 33; and "Gobble Napkins," page 34 courtesy of Andrew Stadel, estimation180.com.

Chapter 5—"Penguins," page 44; "LEGO Bricks" and "Dots," page 47; "Trees," page 49; "Dots 2," page 50; "Dots 3" and "Dots 4," page 51; "Dots 5"and "Dots 6," page 52; and "Circles," page 53 modified courtesy of Fawn Nguyen, visualpatterns.org.

Chapter 5—"Student Work 1," page 49; "Student Work 2," page 50; "Student Work 3," page 50, courtesy of Fawn Nguyen, visualpatterns.org.

Chapter 6—"Shapes," page 56 courtesy of Mary Bourassa, wodb.ca; "Graphs," page 58 courtesy of Chris Hunter, wodb.ca; "Music," page 59 courtesy of Mishaal Surti, wodb.ca

Chapter 7—"Catalonia," page 67, "Colombia," page 69, "Greece," "Thailand," "Bahamas," and "Finland," page 71; "Rectangles," page 72; "Shaded 1" and "Shaded 2," page73; "Shaded 3," page 75; "Shaded 4," page 76; "Shaded 5," "Shaded 6," and "Triangles," page 77 courtesy of Nat Banting, fractiontalks.com.

Chapter 9—"Pumpkins," page 90, CC BY 4.0 by Kristina Peters.

Acknowledgments

I am forever grateful for parents who have modeled perseverance, teamwork, and dedication. Yes, "I wouldn't be here without you" is true, but I also know I wouldn't be in a position to share these words with others without the love, support, and constant encouragement you gave me to chase after a dream. Whether it was taking us to baseball showcases and tournaments, giving up your interests for ours, or sacrificing your needs to ensure our success, you put my brother and me in a position to succeed, and we've become who we are because of it.

Speaking of my brother, you have been the one to always urge and encourage me to take risks. Whether it was driving the Honda Accord through the mud or playing long toss until our arms were ready to fall off, you have always pushed me to be a better version of myself.

No book gets written in a household of two young boys without a caring and understanding spouse. The first time I hung out with my wife was in college and she brought a tin of cookies, a smile, and a demand of being dealt into the game of cards my roommates and I were playing. Your mentality has never changed; you have always taken good care of me, always had a positive attitude, and always supported my crazy ideas with both feet in the boat.

And my two young boys, you are the inspiration for this book and the reason I work so hard. Thank you for unknowingly allowing me to be away from you more than you'd like; I promise I will make it up to you. The life and love you two show for each other and your friends are something I will cherish forever. So many people have reminded us to "enjoy the moments while they last" and you have made this easy. My hope is you stay curious, interested, and passionate about what you believe in.

As you may have realized, I wouldn't have written this book without the work of friends from all over the world. Thank you to the #MTBoS for pushing me to think harder and deeper about how and *why* we teach our students certain topics. I'm privileged to be a part of the community, and I look forward to what is yet to come.

Specifically, I would be remised if I didn't acknowledge and thank Matt Vaudrey, the person who has pushed me to take more professional risks and fine-tune my craft more than anyone I've ever met. Your friendship and encouragement is something I truly appreciate, so thank you.

Finally, thank you to the Burgess family for believing in me enough to bring me into the Dave Burgess Publishing team. I have loved working alongside you as we work to share more ideas with teachers and parents.

More From

DAVE BURGESS
Consulting, Inc.

Teach Like a PIRATE

Increase Student Engagement, Boost Your Creativity, and Transform Your Life as an Educator
By Dave Burgess (@BurgessDave)

Teach Like a PIRATE is the New York Times' best-selling book that has sparked a worldwide educational revolution. It is part inspirational manifesto that ignites passion for the profession, and part practical road map filled with dynamic strategies to dramatically increase student engagement. Translated into multiple languages, its message resonates with educators who want to design outrageously creative lessons and transform school into a life-changing experience for students.

Learn Like a PIRATE

Empower Your Students to Collaborate, Lead, and Succeed

By Paul Solarz (@PaulSolarz)

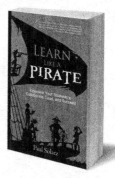

Today's job market demands that students be prepared to take responsibility for their lives and careers. We do them a disservice if we teach them how to earn passing grades without equipping them to take charge of their education. In *Learn Like a PIRATE*, Paul Solarz explains how to design classroom experiences that encourage students to take risks and explore their passions in a stimulating, motivating, and supportive environment where improvement, rather than grades, is the focus. Discover how student-led classrooms help students thrive and develop into self-directed, confident citizens who are capable of making smart, responsible decisions, all on their own.

P is for PIRATE

Inspirational ABC's for Educators

By Dave and Shelley Burgess (@Burgess_Shelley)

Teaching is an adventure that stretches the imagination and calls for creativity every day! In *P is for PIRATE*, husband and wife team Dave and Shelley Burgess encourage and inspire educators to make their classrooms fun and exciting places to learn. Tapping into years of personal experience and drawing on the insights of more than seventy educators, the authors offer a wealth of ideas for making learning and teaching more fulfilling than ever before.

Play Like a Pirate

Engage Students with Toys, Games, and Comics

by Quinn Rollins (@jedikermit)

Yes! School can be simultaneously fun and educational. In *Play Like a Pirate*, Quinn Rollins offers practical, engaging strategies and resources that make it easy to integrate fun into your curriculum. Regardless of the grade level you teach, you'll find inspiration and ideas that will help you engage your students in unforgettable ways.

eXPlore Like a Pirate

Gamification and Game-Inspired Course Design to Engage, Enrich, and Elevate Your Learners

By Michael Matera (@MrMatera)

Are you ready to transform your classroom into an experiential world that flourishes on collaboration and creativity? Then set sail with classroom game designer and educator Michael Matera as he reveals the possibilities and power of game-based learning. In *eXPlore Like a Pirate*, Matera serves as your experienced guide to help you apply the most motivational techniques of gameplay to your classroom. You'll learn gamification strategies that will work with and enhance (rather than replace) your current curriculum and discover how these engaging methods can be applied to any grade level or subject.

Lead Like a PIRATE

Make School Amazing for Your Students and Staff

By Shelley Burgess and Beth Houf
(@Burgess_Shelley, @BethHouf)

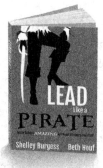

In *Lead Like a PIRATE*, education leaders Shelley Burgess and Beth Houf map out the character traits necessary to captain a school or district. You'll learn where to find the treasure that's already in your classrooms and schools—and how to bring out the very best in your educators. This book will equip and encourage you to be relentless in your quest to make school amazing for your students, staff, parents, and communities.

The Zen Teacher

Creating FOCUS, SIMPLICITY, and TRANQUILITY in the Classroom

By Dan Tricarico (@TheZenTeacher)

Teachers have incredible power to influence—even improve—the future. In *The Zen Teacher*, educator, blogger, and speaker Dan Tricarico provides practical, easy-to-use techniques to help teachers be their best—unrushed and fully focused—so they can maximize their performance and improve their quality of life. In this introductory guide, Dan Tricarico explains what it means to develop a Zen practice—something that has nothing to do with religion and everything to do with your ability to thrive in the classroom.

Master the Media

How Teaching Media Literacy Can Save Our Plugged-in World

By Julie Smith (@julnilsmith)

Written to help teachers and parents educate the next generation, *Master the Media* explains the history, purpose, and messages behind the media. The point isn't to get kids to unplug; it's to help them make informed choices, understand the difference between truth and lies, and discern perception from reality. Critical thinking leads to smarter decisions—and it's why media literacy can save the world.

The Innovator's Mindset

Empower Learning, Unleash Talent, and Lead a Culture of Creativity

By George Couros (@gcouros)

The traditional system of education requires students to hold their questions and compliantly stick to the scheduled curriculum. But our job as educators is to provide new and better opportunities for our students. It's time to recognize that compliance doesn't foster innovation, encourage critical thinking, or inspire creativity—and those are the skills our students need to succeed. In *The Innovator's Mindset*, George Couros encourages teachers and administrators to empower their learners to wonder, to explore—and to become forward-thinking leaders.

50 Things You Can Do with Google Classroom

By Alice Keeler and Libbi Miller
(@AliceKeeler, @MillerLibbi)

It can be challenging to add new technology to the classroom, but it's a must if students are going to be well-equipped for the future. Alice Keeler and Libbi Miller shorten the learning curve by providing a thorough overview of the Google Classroom App. Part of Google Apps for Education (GAfE), Google Classroom was specifically designed to help teachers save time by streamlining the process of going digital. Complete with screenshots, *50 Things You Can Do with Google Classroom* provides ideas and step-by-step instructions to help teachers implement this powerful tool.

50 Things to Go Further with Google Classroom

A Student-Centered Approach

By Alice Keeler and Libbi Miller
(@AliceKeeler, @MillerLibbi)

Today's technology empowers educators to move away from the traditional classroom where teachers lead and students work independently—each doing the same thing. In *50 Things to Go Further with Google Classroom: A Student-Centered Approach*, authors and educators Alice Keeler and Libbi Miller offer inspiration and resources to help you create a digitally rich, engaging, student-centered environment. They show you how to tap into the power of individualized learning that is possible with Google Classroom.

Pure Genius

Building a Culture of Innovation and
Taking 20% Time to the Next Level

By Don Wettrick (@DonWettrick)

For far too long, schools have been bastions of boredom, killers of creativity, and way too comfortable with compliance and conformity. In *Pure Genius*, Don Wettrick explains how collaboration—with experts, students, and other educators—can help you create interesting, and even life-changing, opportunities for learning. Wettrick's book inspires and equips educators with a systematic blueprint for teaching innovation in any school.

140 Twitter Tips for Educators

Get Connected, Grow Your Professional
Learning Network, and Reinvigorate Your Career

By Brad Currie, Billy Krakower, and Scott Rocco
(@bradmcurrie, @wkrakower, @ScottRRocco)

Whatever questions you have about education or about how you can be even better at your job, you'll find ideas, resources, and a vibrant network of professionals ready to help you on Twitter. In *140 Twitter Tips for Educators* #Satchat hosts and founders of Evolving Educators, Brad Currie, Billy Krakower, and Scott Rocco offer step-by-step instructions to help you master the basics of Twitter, build an online following, and become a Twitter rock star.

Ditch That Textbook

Free Your Teaching and Revolutionize
Your Classroom

By Matt Miller (@jmattmiller)

Textbooks are symbols of centuries-old education. They're often outdated as soon as they hit students' desks. Acting "by the textbook" implies compliance and a lack of creativity. It's time to ditch those textbooks—and those textbook assumptions about learning! In *Ditch That Textbook*, teacher and blogger Matt Miller encourages educators to throw out meaningless, pedestrian teaching and learning practices. He empowers them to evolve and improve on old, standard teaching methods. *Ditch That Textbook* is a support system, toolbox, and manifesto to help educators free their teaching and revolutionize their classrooms.

How Much Water Do We Have?

5 Success Principles for Conquering Any Change and Thriving in Times of Change

by Pete Nunweiler with Kris Nunweiler

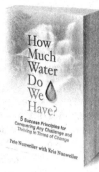

In *How Much Water Do We Have?* Pete Nunweiler identifies five key elements—information, planning, motivation, support, and leadership—that are necessary for the success of any goal, life transition, or challenge. Referring to these elements as the 5 Waters of Success, Pete explains that, like the water we drink, you need them to thrive in today's rapidly paced world. If you're feeling stressed out, overwhelmed, or uncertain at work or at home, pause and look for the signs of dehydration. Learn how to find, acquire, and use the 5 Waters of Success—so you can share them with your team and family members.

Instant Relevance

Using Today's Experiences in Tomorrow's Lessons

By Denis Sheeran (@MathDenisNJ)

Every day, students in schools around the world ask the question, "When am I ever going to use this in real life?" In *Instant Relevance*, author and keynote speaker Denis Sheeran equips you to create engaging lessons *from* experiences and events that matter to your students. Learn how to help your students see meaningful connections between the real world and what they learn in the classroom—because that's when learning sticks.

The Classroom Chef

Sharpen Your Lessons. Season Your Classes. Make Math Meaningful.

By John Stevens and Matt Vaudrey (@Jstevens009, @MrVaudrey)

In *The Classroom Chef*, math teachers and instructional coaches John Stevens and Matt Vaudrey share their secret recipes, ingredients, and tips for serving up lessons that engage students and help them "get" math. You can use these ideas and methods as-is, or better yet, tweak them and create your own enticing educational meals. The message the authors share is that, with imagination and preparation, every teacher can be a Classroom Chef.

Start. Right. Now.

Teach and Lead for Excellence

By Todd Whitaker, Jeff Zoul, and Jimmy Casas
(@ToddWhitaker, @Jeff_Zoul, @casas_jimmy)

In their work leading up to *Start. Right. Now.* Todd Whitaker, Jeff Zoul, and Jimmy Casas studied educators from across the nation and discovered four key behaviors of excellence: Excellent leaders and teachers *Know the Way, Show the Way, Go the Way, and Grow Each Day.* If you are ready to take the first step toward excellence, this motivating book will put you on the right path.

The Writing on the Classroom Wall

How Posting Your Most Passionate Beliefs about Education Can Empower Your Students, Propel Your Growth, and Lead to a Lifetime of Learning

By Steve Wyborney (@SteveWyborney)

In *The Writing on the Classroom Wall*, Steve Wyborney explains how posting and discussing Big Ideas can lead to deeper learning. You'll learn why sharing your ideas will sharpen and refine them. You'll also be encouraged to know that the Big Ideas you share don't have to be profound to make a profound impact on learning. In fact, Steve explains, it's okay if some of your ideas fall *off* the wall. What matters most is sharing them.

LAUNCH

Using Design Thinking to Boost Creativity and Bring Out the Maker in Every Student

By John Spencer and A.J. Juliani
(@spencerideas, @ajjuliani)

Something happens in students when they define themselves as *makers* and *inventors* and *creators*. They discover powerful skills—problem-solving, critical thinking, and imagination—that will help them shape the world's future … *our* future. In *LAUNCH*, John Spencer and A.J. Juliani provide a process that can be incorporated into every class at every grade level … even if you don't consider yourself a "creative teacher." And if you dare to innovate and view creativity as an essential skill, you will empower your students to change the world—starting right now.

Kids Deserve It!

Pushing Boundaries and Challenging Conventional Thinking

By Todd Nesloney and Adam Welcome
(@TechNinjaTodd, @awelcome)

In *Kids Deserve It!*, Todd and Adam encourage you to think big and make learning fun and meaningful for students. Their high-tech, high-touch, and highly engaging practices will inspire you to take risks, shake up the status quo, and be a champion for your students. While you're at it, you just might rediscover why you became an educator in the first place.

Escaping the School Leader's Dunk Tank

How to Prevail When Others Want to See You Drown

By Rebecca Coda and Rick Jetter
(@RebeccaCoda, @RickJetter)

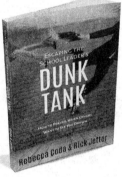

No school leader is immune to the effects of discrimination, bad politics, revenge, or ego-driven coworkers. These kinds of dunk-tank situations can make an educator's life miserable. By sharing real-life stories and insightful research, the authors (who are dunk-tank survivors themselves) equip school leaders with the practical knowledge and emotional tools necessary to survive and, better yet, avoid getting "dunked."

Your School Rocks...So Tell People!

Passionately Pitch and Promote the Positives Happening on Your Campus

By Ryan McLane and Eric Lowe
(@McLane_Ryan, @EricLowe21)

Great things are happening in your school every day. The problem is, no one beyond your school walls knows about them. School principals Ryan McLane and Eric Lowe want to help you get the word out! In *Your School Rocks ... So Tell People!* McLane and Lowe offer more than seventy immediately actionable tips along with easy-to-follow instructions and links to video tutorials. This practical guide will equip you to create an effective and manageable communication strategy using social media tools. Learn how to keep your students' families and community connected, informed, and excited about what's going on in your school.

Teaching Math with Google Apps

50 G Suite Activities

By Alice Keeler and Diana Herrington
(@AliceKeeler, @mathdiana)

Google Apps give teachers the opportunity to interact with students in a more meaningful way than ever before, while G Suite empowers students to be creative, critical thinkers who collaborate as they explore and learn. In *Teaching Math with Google Apps*, educators Alice Keeler and Diana Herrington demonstrate fifty different ways to bring math classes into to the twenty-first century with easy-to-use technology.

About the Author

John Stevens is an educational technology coach for Chaffey Joint Union High School District. He has also taught high school Geometry and Algebra 1, middle school math, and Service Learning, as well as Robotics, Engineering, and Design since 2006. He has served as the go-to guy for trying new, crazy, and often untested ideas to see how well they will work.

He co-founded and moderates #CAedchat, the weekly teacher Twitter chat for the state of California. His site, *Would You Rather?*, is dedicated to getting students talking about math. He is an author of *Flipping 2.0*, a resource to help teachers take the flipped classroom to the next level. He is also the co-author of *The Classroom Chef*, a book about taking risks and building culture in a classroom.

John has had the honor of traveling around the country, sharing his message and risk-taking personality with school districts and at various technology and math conferences. John blogs at fishing4tech. com. His latest adventure has led him into the world of 3D-printing and designing lesson plans and curricula for AirWolf3D.

@jstevens009 or @TableTalkMath

john@tabletalkmath.com

CPSIA information can be obtained
at www.ICGtesting.com
Printed in the USA
FSOW04n0042130917
38457FS